Certified Wireless IoT Design Study Guide
(CWIDP-401)

Copyright © 2022 by CertiTrek Publishing and CWNP, LLC. All rights reserved. Printed in the United States of America. Except as permitted under the United States Copyright Act of 1976, no part of this publication may be reproduced or distributed in any form or by any means, or stored in a data base or retrieval system, without the prior written permission of the publisher.

All trademarks or copyrights mentioned herein are the possession of their respective owners and CertiTrek Publishing makes no claim of ownership by the mention of products that contain these marks.

Errata, when available, for this study guide can be found at: www.cwnp.com/errata/

First printing: September 2021, version 1.0

ISBN: 978-1-7372166-4-3

In addition to the authors of this book, listed in the About the Authors section of the Introduction, CWNP would like to say a special thanks to all those involved in the development of materials for CWIDP-401 from the Job Task Analysis (JTA) through to materials review and feedback. These individuals include Robert Bartz, Ian Beyer, Tom Carpenter, Jonathan Davis, Landon Foster, Manon Lessard, Peter Mackenzie, Troy Martin, Scott McNeil, Phil Morgan, Ferney Munoz, Jim Palmer, and Djamel Ramoul. If we have left out your name it is only because so many helped and not because you were not appreciated. Many thanks to all of you.

Table of Contents

Table of Contents	iii
Introduction	iv
Extended Table of Contents	xi
Chapter 1: Introduction to Wireless IoT Design	1
Chapter 2: Assessing an Existing IoT Solution	39
Chapter 3: Gathering Business Requirements and Constraints	67
Chapter 4: Defining Technical Requirements	137
Chapter 5: IoT Architectures, Topologies, and Protocols	191
Chapter 6: Designing the IoT Wireless Network	237
Chapter 7: Designing the IoT Solution Beyond the Network	347
Chapter 8: Deploying and Validating the IoT Solution	411

Introduction

The Certified Wireless IoT Design Professional (CWIDP) is an individual who can gather requirements, design, validate, and optimize a wireless IoT solution. The CWIDP-401 exam tests the candidate's knowledge to verify his or her ability to perform the duties of a CWIDP.

The Certified Wireless IoT Design Professional (CWIDP) has the knowledge and skill set required to define, design, validate and assess wireless IoT solutions. This professional gathers and defines requirements in collaboration with the appropriate stakeholders in order to design wireless IoT networks and related infrastructure with appropriate security considerations. The CWIDP creates design documentation to support the deployment of the required system components and future operations.

The CWIDP-401 exam consists of 60 multiple choice, single correct answer questions and is delivered through Pearson VUE. The candidate can register for the exam at the Pearson VUE website (https://home.pearsonvue.com/). The candidate will have 90 minutes to take the exam and must achieve a score of 70% or greater to earn the CWIDP certification. If the candidate desires to become a Certified Wireless Network Trainer (CWNT) the passing score must be 80% or greater. A CWNT is authorized to teach official CWNP courses for certifications in which they hold the CWNT credential.

Book Features

The CWIDP Certified Wireless IoT Design Professional Study and Reference Guide includes the following features:

- End of chapter review quizzes. These quizzes help you test the knowledge you've acquired from the chapter. Each chapter contains 10 quiz items.

- Extensive footnotes throughout with more than 200 footnotes providing references, additional information, and author opinions.

- Complete coverage of the CWIDP-401 objectives. Every objective is covered in the book and each chapter lists the objectives covered in part or entirety within.

About the Authors

The following individuals wrote one or more sections of this book:

Darrell DeRosia

Troy Martin

Tom Carpenter

Djamel Ramoul

All these authors are either CWISEs, CWNEs, or experts in the field with more than ten years of experience working with various wireless technologies independently and more than 50 years of experience combined.

At CWNP, we would like to thank all these authors for their hard work and dedication in making this book a reality. They have brought significant value to thousands.

CWIDP-401 Objectives

The CWIDP-401 exam tests your knowledge against four knowledge domains as documented in Table i.2. The CWIDP candidate should understand these domains before taking the exam. The CWIDP-401 objectives follow.

Knowledge Domain	Percentage
Assess an Existing IoT Solution	5%
Gather and Define Requirements and Constraints	30%
Design a Wireless IoT Solution to Meet Requirements	40%
Validate and Optimize the Wireless IoT Solution	25%

Table i.2: CWIDP-404 Exam Knowledge Domains with Percentage of Questions in Each Domain

1.0 Assess an Existing IoT Solution – 5%

1.1 Evaluate an existing IoT implementation and understand its impact on a new wireless IoT deployment

1.2 Use appropriate tools to analyze existing IoT implementations

    1.2.1 Protocol analyzers (wired and wireless)
    1.2.2 Spectrum analyzers
    1.2.3 Network diagrams

1.3 Gather system documentation for the existing IoT solution

1.4 Evaluate operational parameters

    1.4.1 Wireless signal coverage
    1.4.2 Frequencies used
    1.4.3 Functionality (Network servers and services used; Protocols implemented)
    1.4.4 Potential impact on new deployments

1.5 Document findings for use in the design of the new wireless IoT solution

2.0 Gather and Define Requirements and Constraints – 30%

2.1 Gather business requirements and constraints

- 2.1.1 Use cases and justification
- 2.1.2 Identify coverage areas
- 2.1.3 Budget and schedule
- 2.1.4 Architectural and aesthetic constraints
- 2.1.5 Industry and regulatory compliance
- 2.1.6 Data/event collection and control requirements
- 2.1.7 Integration requirements

2.2 Gather technical requirements and constraints

- 2.2.1 Obtain, create, and validate site plans
- 2.2.2 Gather environment characteristics and RF measurements
- 2.2.3 Define device and application data requirements for each area
- 2.2.4 Gather and define system requirements
    - Network topology, capacity, and redundancy
    - Wireless IoT architecture
    - IoT technologies aligned with requirements
    - Location services (geofencing, asset tracking, etc.)
    - Duty cycle, power consumption, and energy harvesting requirements
    - Security requirements
    - Environment conditions
    - Node and tag types and capabilities
    - Device mobility
    - Vendor selection
- 2.2.5 Gather and define operational requirements
    - System monitoring
    - Data collection parameters
    - IoT upgrade requirements, when applicable
- 2.2.6 Gather and define network infrastructure requirements of the planned wireless IoT solution
- 2.2.7 Gather and define cabling infrastructure requirements of the planned wireless IoT solution
- 2.2.8 Document existing wireless systems, designed, and related documentation, when applicable

3.0 Design a Wireless IoT Solution to Meet Requirements – 40%

3.1 Design for selected topologies

- 3.1.1 Mesh
- 3.1.2 PtP
- 3.1.3 PtMP
- 3.1.4 P2P
- 3.1.5 Tree
- 3.1.6 Star
- 3.1.7 Cluster tree

3.2 Design for appropriate channel configuration

- 3.2.1 Channel selection
- 3.2.2 Channel and protocol functionality
- 3.2.3 Blocklist or blocked channels

3.3 Design based on RF requirements and capabilities

- 3.3.1 Use RF measurements and survey tools
- 3.3.2 Use RF modeling tools
- 3.3.3 Perform continuous wave (CW) testing
- 3.3.4 Perform onsite coverage testing/Proof of Concept (PoC)

3.4 Use wireless IoT tools to create and validate the design

- 3.4.1 Generate a predictive RF model using wireless design tools
- 3.4.2 Use additional tools to assist in the design process
- 3.4.3 Utilize validation tools

3.5 Produce or recommend designs and configuration parameters for the IoT-related network infrastructure requirements

- 3.5.1 Required infrastructure hardware and software
- 3.5.2 Required PoE and power budgets
- 3.5.3 Recommend robust security solutions

      3.5.4   Required QoS configuration based on the selected wireless IoT protocol and supported wired network QoS parameters

3.6 Produce design documentation

      3.6.1   Bill of Materials (BoM)
      3.6.2   Design report
      3.6.3   Physical installation guide

4.0 Validate and Optimize the Wireless IoT Solution – 25%

4.1 Validate that the RF requirements are met by the solution

      4.1.1   Ensure coverage requirements are met
      4.1.2   Ensure capacity requirements are met
      4.1.3   Identify and resolve interference sources, when applicable

4.2 Validate the IoT solution is functioning as defined in the solution requirements

      4.2.1   Conduct device testing
      4.2.2   Conduct mobility testing
      4.2.3   Verify proper security configuration and firmware/software support
      4.2.4   Verify proper node (or asset tag) and antenna installation per design specifications and location
      4.2.5   Verify power and grounding requirements are met
      4.2.6   Verify channel selections and transmit power
      4.2.7   Verify aesthetic requirements are met

4.3 Recommend and/or perform appropriate corrective actions as needed based on validation results for RF requirements and IoT solution functionality requirements

4.4 Create a validation and test report including solution documentation and asset inventory/asset documentation

4.5 Final meeting (Q&A and hand-off)

x

Extended Table of Contents

Table of Contents ----- iii

Introduction ----- iv

   Book Features ----- v

   About the Authors ----- vi

   CWIDP-401 Objectives ----- vii

Extended Table of Contents ----- xi

Chapter 1: Introduction to Wireless IoT Design ----- 1

   1.1: The Internet of Things (IoT) Today ----- 2

   1.2: IoT Design is about More than the Network ----- 8

   1.3: The Design Process ----- 11

   1.4: Implementing Effective Project Management ----- 22

   1.5: Keeping Up with the Pace of Change in IoT ----- 27

   1.6: Chapter Summary ----- 32

   1.7: Review Questions ----- 33

   1.8 Review Answers ----- 36

Chapter 2: Assessing an Existing IoT Solution ----- 39

   2.1: What is an Existing System? ----- 40

   2.2: Evaluating Operational Parameters ----- 43

   2.3: Gathering Documentation ----- 48

   2.4: Tools for Analyzing Existing Systems ----- 49

   2.5: Documenting Assessment Results ----- 59

   2.6: Chapter Summary ----- 62

   2.7: Review Questions ----- 63

   2.8: Review Answers ----- 65

Chapter 3: Gathering Business Requirements and Constraints ----- 67

3.1: Requirements Engineering Framework --71

3.2: Defining Stakeholders --77

3.3: Turning Needs into Requirements--79

3.4: Use Cases and Solution Justification ---94

3.5: Coverage Area Requirements--95

3.6: System Users ---97

3.7: System Interfaces and Integration --98

3.8: IoT System Data and Integration--- 102

3.9: Common Constraints --- 103

3.10: Industry and Regulatory Compliance--- 107

3.11: Quality Requirements Specification --- 113

3.12: Writing Requirements: A Case Study --- 124

3.13: Chapter Summary-- 129

3.14: Review Questions -- 131

3.15: Review Answers-- 135

Chapter 4: Defining Technical Requirements--- 137

4.1: Technical Requirements Defined-- 138

4.2: The IEEE 29148-2018 Technical Requirements Process ------------------------- 141

4.3: Exploring Functional and Non-Functional Requirements --------------------------- 149

4.4: Documenting the Requirements--- 177

4.5: Chapter Summary --- 185

4.6: Review Questions-- 186

4.7: Review Answers --- 189

Chapter 5: IoT Architectures, Topologies, and Protocols --------------------------------- 191

5.1: Architecture and Topology Defined--- 192

5.2: Wireless IoT Architectural Reference Frameworks -------------------------------- 198

- 5.3: Creating a Concrete Architecture — 217
- 5.4: Wireless IoT Topologies — 223
- 5.5: Chapter Summary — 232
- 5.6: Review Questions — 233
- 5.7: Review Answers — 236

Chapter 6: Designing the IoT Wireless Network — 237
- 6.1: A Design Framework — 239
- 6.2: Selecting and Building IoT Solutions — 242
- 6.3: Wireless IoT Network Design — 243
- 6.4: Using Wireless Design Software — 299
- 6.5: Wired Network Design/Planning — 313
- 6.6: Supporting Services Design/Planning — 315
- 6.7: Designing Security — 321
- 6.8: Integrating with Existing Industrial Networks — 331
- 6.9: Chapter Summary — 342
- 6.10: Review Questions — 343
- 6.11: Review Answers — 345

Chapter 7: Designing the IoT Solution Beyond the Network — 347
- 7.1: IoT Data Protocol Planning — 348
- 7.2: Data Storage and Streaming Plans — 373
- 7.3: Application Planning — 391
- 7.4: Monitoring and Maintenance Planning — 397
- 7.5: Security above Layers 1-3 — 401
- 7.6: Producing Design Documentation — 403
- 7.7: Concurrent vs. Sequential Design — 404
- 7.7: Chapter Summary — 406

7.8: Review Questions -------- 407

7.9: Review Answers -------- 410

Chapter 8: Deploying and Validating the IoT Solution -------- 411

8.1: Deploying the Solution -------- 412

8.2: Validate RF Requirements -------- 419

8.3: Validate IoT Solution Requirements -------- 421

8.4: Recommend or Perform Corrective Actions in a Validation and Test Report ----- 425

8.5: Final Meeting (Q&A and Hand-Off) -------- 426

8.6: Chapter Summary -------- 427

8.7: Review Questions -------- 428

8.8: Review Answers -------- 430

Chapter 1: Introduction to Wireless IoT Design

Objectives Covered

1.1 Evaluate an existing IoT implementation and understand its impact on a new wireless IoT deployment

The Internet of Things market is growing. Of course, that is an understatement. It is multiplying, extending, expanding, and upsurging. This explosive growth has created a demand for professionals who know how to design IoT solutions and not just IoT devices. The solutions include the network, the devices, and the applications. This knowledge is imperative to becoming a successful Certified Wireless IoT Design Professional (CWDP). *The CWIDP Certified Wireless IoT Design Professional Study and Reference Guide* will help you build the essential skills needed for success. Additionally, the intermediate knowledge you learned in your CWISA studies will be priceless along this journey.

To be an effective wireless IoT solutions designer, experience must be coupled with knowledge, and, in this rapidly expanding market, this knowledge cannot become stale. The professional must learn continually in this space much like networking professionals did in the 1980s and 1990s or like Internet developers did in the 1990s and 2000s. The pace of change in IoT is sometimes staggering.

This chapter introduces several important concepts that will begin this journey. First, IoT is explored as it exists in the early 2020s and then the big picture of IoT design is presented. Next, the design process is summarized, which will be the foundational structure for the remainder of the book. Finally, project management and self-management, as they relate to IoT are addressed. This is a journey well worth taking and it starts with an understanding of IoT as it was and as it is today.

1.1: The Internet of Things (IoT) Today

IoT may be defined as connected things. A more expansive definition would be:

> The Internet of Things (IoT) is the interconnection of things (physical and virtual, mobile and stationary) using connectivity protocols and data transfer protocols that allow for monitoring, sensing, actuation and interaction with and by the things at any time and in any location[1].

---

[1] Bartz, Carpenter, Martin. *Certified Wireless IoT Integration Professional (CWIIP) Study and Reference Guide (CWIIP-301)*. 2021. CertiTrek Publishing

The general mantra of IoT is any thing, any time, and any place. To accomplish this, all layers of the OSI stack must be addressed. The connectivity layers (Layer 1 and Layer 2), the data transport layers (Layer 3 and Layer 4) and the application layers (Layers 5 though 7) must all be incorporated into a well-designed plan. Implementing IoT in the home is one thing. Implementing it in a large-scale commercial, industrial, government, or enterprise setting is completely different. IoT devices do not, of necessity, connect to the Internet[2]. The CWIDP is a professional who can design well-formed plans for wireless IoT solutions that connect to the Internet and those that do not.

Every year, new projections are announced, which are based on surveys of the IoT market, and new technologies are released. To keep up with it all is quite daunting (more on that later in this chapter), but the CWIDP with the right foundations will find it much easier.

2021 IoT Market Studies

In 2021, several studies of the IoT market were released. The Eseye 2021 State of IoT Adoption Report[3] provides the following statistics:

- 500 UK and US senior decision makers and implementers of IoT surveyed.
- 98% said that COVID-19 impacted their IoT projects.
  - 27% said it accelerated development.
  - 31% said they increased investment plans.
  - 21% said it slowed down development.
  - 26% said it led to the cancellation of their IoT project.
- 86% said that IoT is a priority for business and 49% are planning further projects in the next few years.
- 89% are planning to increase budget with 44% planning to boost spending by 51-100%.

---

[2] In fact, the term *Internet* was injected into the name in the 1990s just because it was a buzzword. SOURCE: www.techrepublic.com/article/how-the-term-internet-of-things-was-invented/
[3] iot.eseye.com/iot-adoption-report-2021, accessed July 2021, research survey performed by Opinion Matters on behalf of Eseye, a global IoT connectivity specialist organization. Eseye makes the AnyNet+ SIM, allowing devices to move among networks without switching SIM cards. Their technology is heavily used in cellular-based IoT connectivity

- 10% had deployed between 10,001 devices and 100,000 devices and 2% had deployed more than 100,000 devices (with 500 surveyed, that's far more than 1.5 million devices deployed among just 60 respondents).
- 42% stated that intelligent edge hardware is a top technology driver and 41% stated that LPWAN (Low-Power WAN) technologies are a top technology driver, while only 38% indicated that 5G was a significant driver.

The preceding statistics show that IoT is certainly poised for growth. The following graph illustrates the transition that has occurred in the past few years and the expectations for the near future.

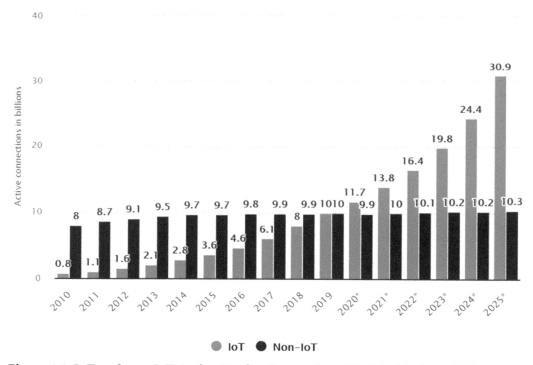

Figure 1.1: IoT and non-IoT Active Device Connections Worldwide from 2010 to 2025

Notice the shift that occurred in 2019. That was the year of balance when IoT and non-IoT devices connected at about the same number (10 billion). From there, the trend is toward growth in IoT connectivity and near stagnation in non-IoT connectivity. The projection is that IoT will total three times the connections of non-IoT by 2025, though some expect the number of connections to be even higher than this[4].

Juniper Research predicts a total of 46 billion connected devices in 2021, which includes IoT in the set and Security Today suggests that 32 billion new IoT devices will be installed around the world.

If you are new to the analysis of all things IoT, these numbers might seem quite large to you. However, consider a single smart city with 900,000 residents would have around 5,000 traffic signals[5], 7,000 fire hydrants, water and sewage treatment plants, temperature, humidity, and noise sensors, and more. The total sensors in a mid-level smart city of this size would easily exceed 20,000 and there are projected to be more than 600 cities larger than 900,000 in population by 2030[6]. Extrapolated out, and only using this low average per city (more than 30 cities will have more than 10 million in population), we can easily see more than 12 billion devices in smart cities alone (17 to 20 billion is a more likely estimate).

Now, consider retail stores. They range from the small food mart at a gas/filling station to mega stores. Many of the larger stores are converting to delivery models where each store is similar to a warehouse that allows people to come in and shop, but its primary purpose is to act as a distribution center for deliveries. Such commercial centers will implement a significant amount of IoT in the coming decade ranging from inventory trackers and deliver drones to, eventually, self-driving vehicles. At a very low estimate of 10,000 IoT devices per store and considering only large mega stores, with more than

---

[4] I predict, based on continual monitoring of research, discussions with professionals in the field, discussions with large organizations and universities, and other analysis that we will see approximately 55-65 (possibly 75) billion IoT devices by 2025 and between 110-120 billion devices by 2030 and these are my conservative estimates. Most studies are too limiting because they have not considered how IoT has changed. Tracking what is "connected to the Internet" is not at all sufficient. What the industry calls IoT is no longer "things connected to the Internet" but instead it is "things connected to anything defined as a network."

[5] The number of traffic signals is typically a factor of city square kilometers and density of streets and population.

[6] The World Cities in 2018 Data Booklet, United Nations

1,000 such stores on average per populated continent, an extremely low estimate of IoT devices is at around 50 billion devices.

These examples should suffice to show that IoT is growing rapidly and will likely continue this growth for the next decade before beginning to slow in the 2030s. But how are we doing at implementing IoT?

IoT Projects and Success

The reality is that many IoT projects either fail outright or fail to reap the desired rewards after implementation. Interestingly, this fact is nothing new to historic technology innovations. We heard similar things about the computer revolution of the 70s and 80s. Then we heard it about the Internet revolution of the 90s and 2000s[7]. Now, it is the reality of the IoT revolution. Anytime a new technology or technological concept is adopted, we see early challenges related to its successful implementation. Part of this is simply because it is new and part of it is because we fail to identify business or organizational drivers and ensure that the planned solution properly links to and provides for them.

Based on research performed by Beecham Research[8] in 2020, only 42% said their IoT projects were fully or mostly successful and 58% said they were mostly unsuccessful or not at all successful. The report finds that the primary causes of IoT project failure were incomplete business aims or objectives, company organizational issues (such as resistance to change), and unforeseen technological problems. An important item to note is that, while 58 felt their projects failed, 49 percent lacked a clear business case. This is like saying, "The project failed, but I don't know why." Without objectives you cannot measure success. Vague business objectives like "increase profits" are difficult to measure in relation to a singular change, such as a new IoT solution. Instead, an objective like "increase profits per product through reduction of manufacturing defects by 23%

---

[7] While many organizations failed to reap value from early Internet-based projects and companies were overvalued creating the "Internet bubble" as it was called, consider the eventual value that came from all these early investments in the Internet. For example, during a global pandemic that required people to stay at home, many companies were able to continue significant levels of operations only because the Internet exists. Today, vast numbers or organizations are so dependent on the Internet, if it were to become unavailable for anything more than a few minutes or hours, the organizations would fail as well. It has become an integral part of business, community, and life for many cultures. The IoT is becoming the same and even more.

[8] www.whyiotprojectsfail.com/

using an IoT solution" is much closer to what is needed. Now we can measure manufacturing defects before and after the implemented IoT solution and determine if we've achieved the objective and, therefore, gained the value desired. Of course, to be most useful, the phrase "using an IoT solution" would be changed to specify a particular IoT solution implemented.

Revisiting the previous Eseye study, 77 percent of respondents said that their IoT project was not as successful as it could have been while only 23 percent said that the project was very successful. The top challenges in these IoT projects were security (39%), device onboarding (35%), testing and certification (35%), and roaming cellular connectivity (35%) for mobile IoT solutions. Future challenges identified included security of the device, connectivity resilience, and uptime.

According to Microsoft research it comes down to the complexity and technical challenges of IoT coupled with a lack of resources and knowledge in most cases[9]. According to The IoT Magazine[10], focusing on a single-issue problem rather than a focus on integration points, lack of understanding of IoT cybersecurity risks, and a lack of skilled professionals are the major pain points.

The things that the research has in common and that the CWIDP can directly address are overcoming technological challenges (including integration and security) and assisting the organization in understanding the complexity and benefits of IoT. Given that the wireless IoT design professional is an expert in IoT design and planning, the individual should be a value-added asset in the planning process. Part of this is assisting the organization in identifying the benefits of the potential IoT solution. Revisiting the Beecham Research, the following are typically considered important IoT project objectives (considered important or very important):

- Improved productivity and efficiencies (100%)
- Reducing costs (100%)
- Increasing safety and security (95%)
- Enabling new business models (94%)
- Improving data and asset utilization (88%)
- Driving revenue (83%)

---

[9] azure.microsoft.com/mediahandler/files/resourcefiles/iot-signals/IoT Signals_Edition 2_English.pdf

[10] theiotmagazine.com/why-do-iot-projects-fail-what-you-must-know-and-do-fa50f4b04833

These drivers can act as a starting point when eliciting business objectives and requirements (discussed in detail in Chapter 3). Of great importance is the concept of enabling new business models. Part of the realization of total benefit in an IoT solution is exploring the integration of various systems within it. This concept is the primary focus of the CWIIP certification and will be considered in this material from a strictly design perspective.

The other area impacted directly by the CWIDP is that of overcoming technological challenges. Finding solutions that meet the needs of the organization within budget is key. Realizing that IoT is about more than just the devices and network is also important (you will read more on this in the next section). The wireless IoT design professional can plan and implement solutions, not just networks. The individual understands the IoT architecture including the connectivity, security, data, and applications used to make it all work together.

With an understanding of the contributors to failure and success, you will see why we have developed the model for wireless IoT solutions design that we have. It is a structured model that ensures requirements are properly identified so that the likelihood of success is greatly increased. Before we explore the design phases, however, it is important to get a basic bird's eye view of what IoT design is all about.

1.2: IoT Design is about More than the Network

A successful wireless IoT solution designer looks beyond the wireless network and understands and plans for the required components to allow for security, integration, and resiliency. This planning requires a solid understanding of IoT architectures and applications, which are covered throughout this book. At this point, it will be helpful to consider IoT solutions that fit the old model of "communicating on the Internet" as well as those that use only local network resources. Many IoT solutions today have no direct Internet use, and they are still considered IoT solutions. Therefore, both must be considered.

IoT with the Internet

The earliest definitions of IoT included the Internet in the definition. Later definitions list the Internet as optional or simply leave it out altogether. For example:

Sensors and actuators embedded in physical objects are linked through wired and wireless networks, often using the same Internet Protocol (IP) that connects the Internet.
-McKinsey Definition

The connection of systems and devices with primarily physical purposes (e.g., sensing, heating/cooling, lighting, motor actuation, transportation) to information networks (including the Internet) via interoperable protocols, often built into embedded systems.
-DHS Definition

A global infrastructure for the information society, enabling advanced services by interconnecting (physical and virtual) things based on existing and evolving, interoperable information and communication technologies.
-ITU Definition

From these definitions it is clear that IoT can use or ignore the Internet and still be called the Internet of Things. It is best, therefore, to think of the Internet, in the Internet of Things, as a generic reference to internetworking technologies rather than that global network we all use to experience the pain of social media and the joy of shopping.

When used with the global Internet, IoT solutions most often take advantage of the Internet for transport or for cloud services. Used only for transport, the Internet effectively acts as a WAN connection. Used for cloud services, the Internet becomes the access method used to transfer data to and from the cloud and to access compute resources in the cloud. Figure 1.2 illustrates the use of the Internet with IoT when cloud technologies are involved.

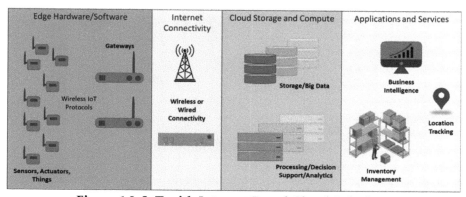

Figure 1.2: IoT with Internet-Based Cloud Solutions

Certainly, IoT architectures can be more complex than that represented in Figure 1.2. For example, the addition of edge computing, fog computing, and complex hybrid cloud solutions would change things. But, for now, it illustrates the general concept required. The edge IoT devices ultimately send their data to the cloud and receive instructions and updates from the cloud. Users access the cloud data with applications (that may run locally or in the cloud) to analyze, process, or otherwise act upon the data. The wireless IoT solutions designer must factor in all these architectural areas when creating a complete design solution.

IoT without the Internet

When used on premises with no Internet data transfer, or possibly using the Internet only for data transfer as a basic WAN solution, the high-level architecture looks like that in Figure 1.3. Notice that the only changes are that we are not using the cloud and we may be limiting communications to the LAN. Instead of using cloud service providers for data storage and compute, we provide all services on the local network. Applications and services access the local resources rather than the cloud. The fundamental point of difference is in where the data is[11] initially stored and where it is processed as well as the added complexity of building the on-premises solutions. When using a cloud service, it's often as simple as enabling a feature and configuring its options. When using local services, it involves deploying the hardware and the software and then configuring its options.

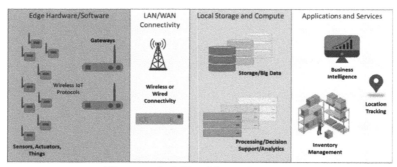

Figure 1.3: IoT without Internet

---

[11] While I am aware that *data* is a plural noun and the *official* use is to say or write "data are" rather than "data is", I have chosen to use "data is" for readability. Additionally, the book most often references data as a collective set, being the singular set of datum, and not some unknown volume of datum. Therefore, the singular verbs simply feel better.

Now that you have the big picture of the system components that might be involved, you can explore the design process CWNP recommends for wireless IoT solution design. It begins with the essential phase of requirements definition, without which the IoT project cannot rationally be considered a success or failure.

1.3: The Design Process

The basic design process used in wireless IoT solution design is explained in this section. It is the same process CWNP recommends for wireless LAN design but involves different actions throughout the process tailored to the needs of IoT solutions. Additionally, we are designing or planning the entire wireless IoT solution from the wireless end devices through to the applications on the network for user and business consumption. However, we are not *building* or *deploying* everything. Therefore, the complex details of things like MQTT, database systems, and such are reserved for the CWIIP certification materials. Here, in later chapters, they will be addressed with sufficient information to place them properly in your design.

The process can be explained in four phases with tasks performed in each phase. The complexity of IoT solution projects all but ensures that some recursion will occur among the first three phases. For example, you may pass through the Define phase, where business requirements and stakeholder requirements are initially developed, only to encounter unexpected issues in the Design phase requiring a revisit of the business and stakeholder requirements. In fact, it is also possible that some requirements engineering will still occur in the Deploy phase of the process. As you will see in chapters 3 and 4, where requirements engineering is explained in detail, this recursion is a normal and expected part of systems engineering, including wireless IoT systems. Therefore, you should not think of the four phases of the design process as purely sequential. The design process phases[12], the recursive nature of requirements engineering, and the flow of a wireless IoT solution design are illustrated in Figure 1.4.

---

[12] You can think of these four phases as sub-phases within the PMI PMBOK Executing and Monitoring and Controlling phases of project management. When we use the term project, we are typically referencing just the technical phases of the project in these higher level PMBOK phases; however, context should reveal the intention.

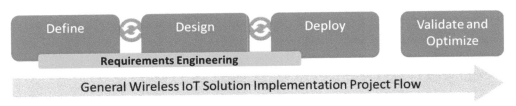

Figure 1.4: The WLAN Design Process

The design process includes four phases: Define, Design, Deploy, and Validate and Optimize. In fact, the knowledge domains of the CWIDP exam follow these four phases (while limiting coverage of "Deploy" as it is addressed in CWICP and CWIIP): Gather and Define Requirements and Constraints (Assess an Existing IoT Solution), Design a Wireless IoT Solution to Meet Requirements, and Validate and Optimize the Wireless IoT Solution. The remaining chapters of the book explore the various tasks performed within this high-level design process in detail and the following sections provide a foundational overview of them.

Define

In the Define phase of a wireless IoT solution design, you will gather important information leading to a specific set of requirements that the design must fulfill. These requirements will include business, stakeholder, and technical requirements. Here are some important tasks to implement when in the Define phase:

- **Assess existing IoT solutions** is evaluating the IoT solutions already in the organization or space where your wireless solution will be implemented and is important. It can help to minimize conflicts between the systems and ensure proper reuse. For example, there is no need to plan an MQTT server or multiple server deployments if one already exists with sufficient capacity and capabilities to handle the new solution. This task is discussed in detail in Chapter 2.

- **Information gathering** is the typical first step in the Define phase of the design process. If a project charter or request for proposal (RFP) exists, this is the first piece of information used and that on which is built all future information and documentation. As a wireless IoT solutions designer with expertise in IoT design, it is your responsibility to communicate with the business stakeholders concerning technical capabilities of a solution. If the project charter or RFP makes requests that are impossible or impractical, this should be communicated to the stakeholders with options for alternate solutions. Starting on a proper, realistic foundation is key to project success.

- **Requirements engineering** is essential to the definition of a any system and, as we've seen in previous sections, it is extremely important for IoT solutions. This stage of defining a solution can consists of multiple tasks. The customer or acquirer, whether internal or external, submits a request for an IoT solution to be designed and installed and the right information must be gathered to determine the type of installation and the equipment and services needed for the installation. All business and stakeholder requirements will be documented in the Define phase. In some cases, some technical requirements will be developed in the Design phase, but most should be discovered in the Define phase as well. Another way of thinking of this is that issues arising in the Design phase may cause recursion back to the Define phase for further requirements specification.

   In addition to requirements, constraints should be identified. With well-formed requirement statements, any applicable constraints are included in the requirement. For example, a business requirement should not state that, "An IoT solution shall be implemented that allows for environmental monitoring within the warehouse to improve employee safety and health." Instead, it should state, "An IoT solution shall be implemented that allows for environmental monitoring within the warehouse to improve employee safety and health within ten months and in compliance with safety regulations." In the latter requirement statement, the constraints are included. Some constraints, such as the overall budget, may be included at the project level or for each requirement, depending on the level of detail.

- **Documenting the requirements** is just as important as generating the requirements. Requirements engineering includes documentation and the process is not complete or accurate without the documentation. Chapters 3 and 4 provide more information on this important aspect of the define phase.

A designer must be skilled in several areas to be successful in performing the tasks required in the define phase. These areas include:

- **Communication Skills:** We cannot teach communication skills here as it would require a book-length treatment of its own, but these skills are essential to success for an IoT solutions designer. Listening is vitally important in the beginning and throughout the design phases to ensure that you understand the needs of the acquirer[13]. Verbal communications are important when communicating technical capabilities and constraints of a solution. Written communications are important for documentation and reporting. Ultimately, communications comprise the foundation for success in this role.

- **Wireless and IoT Knowledge:** The wireless IoT solutions designer must know the wireless protocols exceptionally well. Designing an RF-based solution without understanding how it works is a recipe for disaster. Additionally, the designer should have an intermediate knowledge of data storage, application protocols, and IoT devices (sensors and their components, and actuators and their components). Much of this knowledge is presented in this book and foundational knowledge is provided in the CWISA materials. In-depth knowledge of data storage and application protocols, as well as scripting and automation, are addressed in the CWIIP materials.

- **Wired LAN Knowledge:** It is not enough to be skilled in the wireless IoT protocols. The wireless IoT solutions designer must also have a strong understanding of wired LAN technologies as well because many wireless IoT solutions ultimately integrate with and use the wired LAN for local data transfer. Knowledge of wired LANs will be of great benefit as you plan for network connections, mobility solutions, security configurations, and other required capabilities of the IoT solution. Beneficial knowledge includes switching and routing, network services, IP addressing, and network segmentation through VLANs, firewalls, and access control lists (ACLs), and more.

- **Cabling Infrastructure Knowledge:** It is beneficial to know fiber optic and cabling standards so that the wireless IoT solution design does not fail when

---

[13] The *acquirer* is the individual or organization authorizing and funding the wireless IoT solution.

gateways are installed, and cable runs are too far from the nearest available switch location. This often becomes a problem for new designers and overlooking the need for a new switch installation, due to insufficient available ports or Power over Ethernet (PoE) power budgets, can become expensive as well as time-consuming. Understanding the wired cabling and power provisioning is important for gateway[14] deployment in most installations. The exception is when the gateway is connected wirelessly itself to a remote network through 4G/LTE, 5G, or some other wireless WAN solution.

- **Customer feedback and agreement:** This area is important as a follow-up after all the requirements engineering and information gathering is complete. Documenting your findings and creating a presentation to support your ideas and suggestions for the customer's design is an important stage before beginning the actual detailed design. You should explain your proposed design concept based on the information gathered from the customer and receive feedback from them to ensure that you have accurately defined the need and that both parties agree.

After the need is clearly defined as requirements, documented, and approved, the Design phase can begin.

Design

The Design phase of the wireless IoT solution design process is where the technical expertise of the designer is used to translate the requirements into technical solutions and to generate the wireless (RF) designs as well as the supporting services designs (IoT higher layer protocols (IoT Data protocols), data storage, integration processes, etc.). Many tasks will be performed in the Design phase and may include:

- Performing a spectrum analysis of the facility (performed pre-design)
- Performing a survey to determine gateway placement (performed pre-design)

---

[14] Throughout this text, the term *gateway* is used generically to reference the device through which the end devices wirelessly connect. In some wireless protocols, this device is called a coordinator, a router, a base station, or a gateway. In the context of a specific protocol that uses the term gateway, we will use the full protocol name and the term gateway, such as LoRaWAN gateway, to differentiate it from the generic concept.

- Making proper equipment selections to support the customer's needs
- Creating design documents using wireless network design software
- Defining equipment configurations
- Designing or recommending the IoT solution supporting services
- Creating a Bill of Materials (BoM)

The purpose of the spectrum analysis is to determine if there are interferers in the area, existing or rogue wireless equipment or unintentional radiators present, or any other issues that may cause limitations on band and channel selection, which impacts the wireless IoT protocol choice and configuration. Proper use of spectrum analysis tools can save a lot of time during the Design phase and can also save a lot of money in person-hours and cable moves caused by re-designing around issues found after the deployment.

The survey is performed by measuring the RF signal propagation (based on necessary power levels and channel settings) of a gateway throughout the facility while knowing the capabilities of the end devices that must communicate with the gateway. Its purpose is to determine correct placement, cell size, and overlapping coverage of each gateway needed to meet the requirements of the new wireless IoT network, which were determined during the Define phase. During the survey, proper gateway model selections are made based on the environment being surveyed with the appropriate antennas and backhaul network connectivity (wired vs. wireless). If needed, additional network switches and locations will be selected along with backup power and security mechanisms. Recognizing the need for other network support items is important to getting a design done right the first time without repeat visits and surprise add-on costs.

In addition, during the survey, identifying the target locations of end devices is important. The location of the end device and the physical entity in which it is embedded or to which it will be attached will determine several important items:

- How will the device be powered? If the location does not provide AC power, the device may be powered by batteries alone or batteries coupled with energy harvesting. In some rare cases, energy harvesting alone may be used (the device becomes active and communicates on the network only when energy is actively harvested and powers the device). An example of the latter would be a push button type device wherein pushing the button generates sufficient energy to quickly power a transmission of a basic signal from the device to the gateway.

- How does the mounting of the device impact transmission and reception? If the device is mounted within a metal enclosure, it will significantly impact its ability to communicate. Using an antenna external to the enclosure may be required for such devices.
- Will the device be mobile or stationary? Mobile end devices will often increase the cell size requirements or demand more gateways be installed providing multiple cells. Alternatively, they may demand a cellular-based solution to take advantage of the existing coverage provided by a service provider.

While additional issues may surface, you can see that an understanding of the location and mounting of end devices is critical in implementing an appropriate wireless IoT solution.

The Design phase is broad and requires significant knowledge and skill in multiple areas of networking and applications. There are times when the RF coverage is the only portion of the network a wireless IoT solution designer will be required to provide, but it is certainly not common, and more is expected from wireless IoT professionals every day. It is best that a wireless IoT solutions professional continually learn in the areas of wireless networking, application architectures, network services, and other key areas[15].

It is important to understand the significance of an accurate survey. Equipment and software cannot make up for bad RF surveys in wireless IoT solution design, nor should they be expected to do so. Take the time to learn the proper techniques required to provide accuracy in this stage to avoid problems with the wireless network after it is in operation.

---

[15] For the entire 30 years of my career in IT I have focused on information as my foundation. By this I mean that I continually remind myself that the "I" in IT standard for information. I need to understand how information is generated, transmitted, stored, secured, analyzed, and destroyed to be a most effective professional. The same is true in the OT (Operations Technology) world today. Rare is the OT professional who is not continually making decisions, taking actions, or planning changes based on information provided by the industrial systems used. Knowing the life cycle of information and the technologies used to make that life cycle work is essential. For this reason, I read a book, take a class, attend a conference, or work with a technology (in the lab at the very least) in all areas of the information flow each year. I do my best to go deep into at least 2-3 client or end devices and how they work, a new network protocol and how it works, a database solution and how it works, and several applications (scripting, analysis, simulation, etc.) each year. This plan keeps my knowledge fresh and my value high and I recommend it to all professionals.

Deploy

While the Deploy phase is not addressed in detail in this material, as stated previously, it is covered in the other three exam-based certification in the CWNP wireless IoT track: CWISA, CWICP, and CWIIP in relation to the different components of an IoT solution. However, the wireless IoT solutions designer should be aware of important considerations for the Deploy phase of the project, which are covered here.

Documentation containing a bill of materials (BoM) is usually provided with a complete design package, and once the design is completed and approval is granted, the equipment is ordered for installation. In wireless IoT projects, it is not uncommon to initially order and install the required gateways and build the required supporting services out without implementing the intended full set of end devices. While the solution may eventually include more than one thousand end devices, the initial implementation may deploy only a few dozen or some other smaller number. This reality means that the deployment must be such that it can support the eventual implementation of many more IoT end devices than those used immediately after installation. The designer should communicate with the acquirer to understand if they desire to install the supporting infrastructure and services for the long-term plan or only that which is sufficient for the near-term installation. Then, the designer can design the full-scale infrastructure if required, or design the near-term solution with allowances for future growth that will not prevent such growth and will not allow the future growth to impede on the functionality of the near-term solution.

In a design plan, a *Bill of Materials (BoM)* is a listing of the hardware and possibly software required to deploy the designed wireless IoT solution. This will include everything required, like antennas, cables, mounting kits, poles, lightning arrestors, end devices, application servers, licenses, and anything else needed to deploy a proper solution.

A wireless IoT solution deployment plan must consider, at least, two high-level architectural areas that must exist for the solution to work:

- **Edge networking:** This part of the deployment plan invokes actions resulting in the existence of proper gateways with proper configurations to provide coverage and capacity for the IoT end devices. It involved configuring the solution for device onboarding and security of the radio link between the end devices and the gateway(s). If the end devices exist but cannot communicate with the gateway, the solution has failed. It is important that the installation technicians follow the plans for the deployment, assuming the design is proper, so that all devices can connect when and as required.

- **Supporting services:** Supporting services may be further subdivided into at least three categories:

  - **IoT Data Protocols:** These protocols include solutions like MQTT, DDS (Data Distribution Service), and AMQP (Advanced Message Queuing Protocol). These protocols are sometimes called machine-to-machine (M2M) protocols or simply IoT protocols[16]. Having the appropriate servers in place allows for the implementation of these protocols, when servers are required. In cloud-based solutions, the same or different protocols, sometimes proprietary, may be used. In such cases, the vendor's gateways will form a connection with the cloud and transfer data from the edge devices to the cloud. The deployment must simply ensure a stable and sufficient Internet connection in the latter case.

  - **Data Storage:** Once the data is passed through some IoT Data protocol, it is either processed on-the-fly (real-time data processing) or stored for later processing or both. In most cases, at least some of the data must be stored somewhere. The storage can be on-premises, in the cloud, or both. Depending on the size of the data set generated over time, the storage solution must be able to accommodate both the ultimate size of the data set and the rate at which it arrives without loss. Therefore, deployment of on-premises data storage may be as simple as installing a MySQL, mongoDB, or SQL Server

---

[16] This is where much confusion often arises in wireless IoT. You have the "wireless IoT protocols," like 6LoWPAN, ISA100.11a, WirelessHART, Zigbee, etc., and then you have the "IoT protocols," like MQTT, DDS, and AMQP, among others. We have chosen to reference the latter as IoT Data protocols for simplicity. Realize, however, that some solutions, such as Zigbee, have elements of IoT Data protocols built into the protocol stack as defined. Solutions also exist to integrate Zigbee with MQTT, however, such as Zigbee2MQTT.

database server (as examples) or as complex as implementing a large-scale Hadoop data cluster with data nodes, name nodes, and redundant network connections. Deployment of cloud storage may be as simple as dumping data into an AWS S3 storage bucket or as complex as implementing an AWS Redshift database solution (as examples). As the designer, you should provide instructions for the level of storage space and resiliency required and allow the experts in those areas to deploy based on your higher-level specifications.

- **Processing:** Where do we process our IoT data? At the edge? On-premises? In the fog? In the cloud? On Mars? Well, that last option may not be available for a few years, but this is an area of key importance in both design and deployment. In deployment, it means provisioning the required processing power where required. In design, it's about understanding the IoT solution and the requirements of real-world interactions that may demand one (or more) of the four (edge, on-premises, fog, cloud) in one case and another of the four in another case. For each such case, the processing capabilities and location should be sufficiently defined in the design for proper deployment.

Many little details of the specific actions taken in deployment should be handled by those with expertise in those areas. For example, the designer needn't specify that the Ethernet cable be functional. That should be assumed. The designer also needn't be an expert in all areas impacting the wireless IoT solution design. Instead, the CWIDP should rely on database systems experts and compute processing experts (for example Artificial Intelligence (AI) and its subsets such as Machine Learning (ML) and Neural Networks). It is impractical to think that one person will have complete expertise in all areas touching IoT today and in the future.

Validate and Optimize

The final phase of a wireless IoT solution implementation project is the Validate and Optimize phase. Here, the designer will either perform validation tasks or have them performed so that the results can be analyzed. The purpose of the validation or post-installation survey process is simple: to ensure that the deployed solution meets the requirements. At a minimum, the following four items should be verified:

- Coverage requirements are met on the wireless side
- Application requirements are met within the system
- All required functional abilities exist
- All quality expectations have been met

Required coverage means sufficient wireless coverage (typically measured by signal strength and connectivity) throughout all the areas listed in the defined requirements. Application requirements indicate that the solution is providing the information to the right location in the right format for consumption by users. Functional abilities ensure that the users can do what they need to do, which is dependent on the data being in the right location and the provisioning of tools with the appropriate capabilities. Quality expectations are related to network-level factors, such as packet loss, data rates, security, etc., and application-level factors such as accurate data, timely data, and security.

This phase can become repetitive if multiple changes are needed to bring the solution into compliance with the customer's requirements. You may test an area and find that adjustments need to be made and then go back and test the area again after you make those adjustments only to find that more corrections are needed. This is known to happen several times before a validation or post-installation survey is complete and, based on the size of the IoT solution, it could carry on for several days before you are confident the solution is at its optimal performance. It is a necessary step and should always be taken seriously because the result of failing to implement the post-installation survey and validation can be poor IoT solution performance or lack of required capability.

With a high-level understanding of the wireless IoT solution design process, you are prepared to explore the remainder of the book. However, before leaving this chapter, it's important to consider general project management concepts in relation to your project and to create a plan for keeping up with the rate of change within the IoT space. These two topics comprise the remainder of this chapter.

1.4: Implementing Effective Project Management

In many cases, the person designing the wireless IoT solution will also have to perform some, or all the project management tasks. With this as a given, it is a good practice to familiarize yourself with the project management process. CWNP does not expect you to be a Project Management Professional (PMP)[17] certified individual to pass the CWIDP exam, but you should know basic concepts related to managing WLAN design and installation projects.

All projects usually start with an idea for a deliverable. The deliverable, in this case, is a wireless IoT solution. The idea is pitched to stakeholders as a business use case, and if it is agreed upon, a project charter is drawn up (hopefully) and signed by all necessary parties; at this point, you have an official project. The project charter[18] is a document that officially recognizes the project, and this book will not go into a lot of detail covering a project charter, since you only need to be familiar enough to understand what it is for. Figure 1.5 shows a portion of a project charter document.

Many companies will require a non-disclosure agreement (NDA) to be signed by you and any other personnel taking part in the project if you are acting as an outside consultant or contractor. This document simply signifies that a confidential relationship between parties has been established and all information about the company and its project that is shared or learned during the project must remain confidential between the parties. This is very common in the corporate world to protect information that could damage the corporation if others accessed it outside the company or competitors were able to take advantage of it.

Once you have a project, you are in the Initiation phase of the PMI model where the project objectives are solidified, the business goals are clarified, and the scope of work is fully defined. You are now ready for project planning. Some of this phase of the project lifecycle is overlaps with the defining phase of the wireless IoT solutions design process.

---

[17] In addition to the PMP certification, CompTIA offers the Project+ certification, which is a less intensive certification process, but it teaches the principles and concepts important for most project managers. Interestingly, it started out as IT Project+ many years ago and was simply rebranded as Project+ along the way.

[18] According to the Project Management Institute, a *project charter* is a document issued by the project initiator or sponsor that formally authorizes the existence of a project and provides a project manager with the authority to apply organizational resources to project activities.

A customer provides you with a business goal and reason for the project, and then you, the designer, perform requirements engineering actions that gather information vital to developing the scope of work.

PROJECT CHARTER

1.0 PROJECT IDENTIFICATION

Name	IoT Environmental Health Monitoring Project	
Description	Design, deploy and validate an IoT health monitoring solution in the company warehouse	
Sponsor	John Smith	
Project Manager	Dale Carpenter	
Project Team Resources	IT Department, OT Department, Legal	Communications, Policy, Healthy Workplace Advisory group, Attraction & Retention working group

2.0 BUSINESS REASONS FOR PROJECT
- Comply with regulations
- Improve worker safety
- Reduce injury and health-related costs

3.0 PROJECT OBJECTIVES (PURPOSE)
- Monitor the environmental health quality in the warehouse by monitoring key metrics and ensuring compliance with regulations
- Provide actionable data for the improvement of air quality and other environmental factors
- Provide automated remediation systems for continuous healthy environment conditions

4.0 PROJECT SCOPE
- Limited to the Marysville, Ohio warehouse
- Inclusive of temperature, humidity, foreign/dangerous particles, and oxygen levels

5.0 KEY PROJECT DELIVERABLES

Name	Description
Project charter	This document
High-Level Architecture	Provides an overview of the systems and networks required for the solution
Requirements Specification	Defines the business, stakeholder, and technical requirements of the IoT solution
Design Plan	Provides detailed specification for the wireless network and supporting services required for functional and quality requirements and inclusive of a Bill of Materials (BOM)
Deployment Schedule	Defines the schedule of installation in the proper sequence for effective deployment
Validation and Optimization Report	Describes the actions taken and the results discovered as well as any remediation actions required to bring the implemented solution into compliance with requirements

Figure 1.5: Wireless IoT Solution Project Charter

The scope of work or statement of work (SoW) defines the work to be done, and it often represents a high-level scope initially but then develops into a more detailed scope as the project progresses. This is not always the case, but the agile approach to large corporate projects is now very common, as changes could develop during many stages of the project that can affect time and cost. The project manager, in partnership with technical experts, will develop a work breakdown structure (WBS) or similar document that will divide the project into multiple tasks that can be assigned to various individuals along with task costs and times to complete. An example of work breakdown would look like this at a high level:

1. Survey the wireless installation location
2. Install fiber for new switches
3. Install cabling for new gateways
4. Install new gateways
5. Test the wireless installation

This is a simple list for demonstration purposes only and is not exhaustive, in fact, steps are not listed here to save space, but a full WBS can become much larger than this. In fact, each of the list items can now be broken down into smaller tasks:

1. **Survey the wireless installation location**

    1.1 Perform spectrum analysis.

    1.2 Locate all existing network switches.

    1.3 Document existing fiber.

    1.4 Document existing wireless networks or radiations (if any).

2. **Install fiber for new switches**

    2.1 Obtain drilling permits.

    2.2 Hang and mount conduit.

    2.3 Rack-mount fiber fan-out boxes.

    2.4 Pull fiber through conduit.

    2.5 Terminate fiber.

    2.6 Test fiber.

3. **Install cabling for new gateways**

    3.1 Locate and identify cabling routes.

    3.2 Install conduit (if needed).

    3.3 Pull and label cable drops.

    3.4 Rack-mount cabling patch panels.

    3.5 Dress and terminate cables.

    3.6 Test cables.

4. **Install new gateways**

    4.1 Configure and label new gateways.

    4.2 Document mac addresses and serial numbers of new gateways.

    4.3 Test new gateway configurations.

    4.4 Mount/physically install new gateways.

5. **Test the wireless installation**

    5.1 Perform an RF coverage test.

    5.2 Test end device connectivity.

    5.3 Test data rate requirements.

    5.4 Test capacity requirements.

    5.5 Test capabilities requirements.

Again, the preceding WBS is only an example and not meant to be used as a real-world WBS, because every project is different and should be managed on an individual basis. The WBS is the "list" that drives your project and acts as a checklist for you along the way. According to *Gawande, Atul, The Checklist Manifesto, 2001*, "Good checklists, on the other hand, are precise. They are efficient, to the point, and easy to use even in the most difficult situations. They do not try to spell out everything – a checklist cannot fly a plane. Instead, they provide reminders of only the most critical and important steps – the ones that even the highly skilled professionals using them could miss. Good checklists are, above all, practical." The same is true for a WBS. It is the work checklist for your wireless IoT solution design and deployment project and should be useful rather than simply showing how much you know about design.

With this WBS example, you can see that the phases of the project can be broken down into smaller individual tasks, and these tasks can be assigned to different individuals to be responsible for them. With this information, a project team is formed with competent employees and contractors who possess the proper skill set to complete the work listed. The WBS helps with time, cost, and resource management[19], as each task can be given a cost and time to complete from the responsible individual assigned to the task.

To help with resource management, a communications matrix[20] could be developed. This document will detail the roles and responsibilities of each team member, how documentation should be distributed to and from each team member, and how each team member and stakeholder prefers to be contacted. It will also list the contact information of each person.

At this point, we have a project plan, a breakdown into individual tasks of the work to be done, competent and responsible team members assigned to each task, a cost and time frame assigned to each task, and a document providing detailed communication methods for each person involved in the project as well as their roles and responsibilities. With all of this in place, you or the project manager can effectively manage the project as it progresses while keeping track of your resources[21].

---

[19] According to the Project Management Institute, a project resource can be skilled human resources (specific disciplines either individually or in crews or teams), equipment, services, supplies, commodities, material, budgets, or funds.

[20] A communications matrix is a document containing the communications management plan for the project. If should list all required contact information and the kind of reporting required to and from the various contacts.

[21] I've spent a significant part of my life studying project management. Early on in my career, I had to manage projects impacting thousands of clients on a national network and quickly saw the need for strong project management skills. Having studied many experts on the topic, I can highly recommend anything written by James P. Lewis in relation to project management. While the books are not focused on technology projects specifically, they provide a wealth of valuable information for the IT and OT professional. While he has authored several books on project management over the years, if you can get just one, I would recommend *Project Planning, Scheduling & Control, Fourth Edition,* and then *Project Leadership.* You can begin the journey learning from him on YouTube here: youtu.be/4v_KQnemu2M (Achieve Excellence in Project Management, Part 1). Be sure to continue to parts 2 and 3 as well. If you haven't been exposed to experienced thinking on project management before, this series will provide a great foundation.

1.5: Keeping Up with the Pace of Change in IoT

IoT is a fast-moving technology space. New protocols are developed annually, and new uses are being discovered almost daily. To keep up with the pace of change and new capabilities, the professional requires a solid plan. We want to suggest four sources of information that can help you stay ahead of the game:

- Standards bodies
- Industry magazines (many free)
- Social media
- Tom's top five YouTube channels

Standards Bodies

The first source of information is the standards bodies. These include the IEEE, IETF, ISO/IEC, and specific protocol groups, of which you became familiar in your CWISA studies. These groups provide standards for communications systems and for various functions of those systems, such as onboarding, security, and management.

Important standards from the IEEE related to IoT include:

- 802.3 – Ethernet
- 802.11 – Wi-Fi (Wireless LAN MAC and PHY)
- 802.15.4 – Low-Rate Wireless Networks (LRWNs)
- 802.15.6 – Wireless Body Area Networks (WBANs)
- 29148-2018 – Systems and Software Engineering – Life Cycle Processes – Requirements Engineering

Important standards and recommendations from the IETF related to IoT include, but certainly are not limited to:

- RFC 4944 – 6LoWPAN (enhanced by RFCs 6282, 6775, 8025, 8066, and 8931)
- RFC 8180 – 6TiSCH (Minimal IPv6 over TSCH[22] Mode of 802.15.4) Configuration
- RFC 8576 – IoT Security – State of the Art and Challenges
- RFC 7668 – IPv6 over Bluetooth Low Energy (BLE)[23]

---

[22] Time-Slotted Channel Hopping (TSCH) is a frequency use method defined in 802.15.4.

Of course, the IETF defined IPv4 and IPv6, which are important as well.

Many other standards could be listed, but an important part of your role will be to seek out and review standards and standards organizations relevant to wireless IoT solutions.

Industry Magazines

By industry magazines, we do not mean, necessarily, those specifically on the topic of IoT, but those that focus on technology within the industries important to you. Specific IoT journals and magazines are also useful. Here are a few to explore, starting with IoT specific journals and magazines:

- IEEE Internet of Things Journal
- IEEE Internet of Things Magazine
- IEEE Sensors Journal
- IEEE Sensors Letters
- Sensors (Open Access, MDPI)
- Automation (Open Access, MDPI)
- Buildings (Open Access, MDPI)
- Electronics (Open Access, MDPI)
- Future Internet (Open Access, MDPI)
- IoT (Open Access, MDPI)
- Journal of Cybersecurity and Privacy (Open Access, MDPI)
- Journal of Sensor and Actuator Networks (Open Access, MDPI)
- Network (Open Access, MDPI)
- Signals (Open Access, MDPI)
- Smart Cities (Open Access, MDPI)
- Telecom (Open Access, MDPI)
- Control Engineering (www.controleng.com/magazine)
- IoT Now (www.iot-now.com/iot-magazine-digital-edition/)
- Automation World (www.automationworld.com)
- Cabling Installation & Maintenance (www.cablinginstall.com/magazine)
- Chemical Processing (www.chemicalprocessing.com/issues/)
- Control Design (www.controldesign.com/issues/)

---

[23] The 6Lo working group also has drafts for IPv6 over Near Field Communication and PLC Networks, as well as proposed standards for ITU-T G.9959 networks.

- COTS Journal of Military Electronics & Computing (www.cotsjournalonline.com/)
- Design Engineering (www.design-engineering.com/digital-edition/)
- Design World (www.designworldonline.com/category/digital-issues/digital-issues-ee/)
- Electronic Products & Technology (www.ept.ca/digital-edition/)
- Electronics Maker (electronicsmaker.com/magazine)
- Essential Install (essentialinstall.com/latest-issue/)
- Eureka! (www.eurekamagazine.co.uk/design-engineering-magazine/)
- Food Processing (www.foodprocessing.com/issues/)
- Healthcare Radius (www.healthcareradius.in/emagazines)
- Instrumentation & Control (www.instrumentation.co.za/archives.aspx)
- Manufacturing Automation (www.automationmag.com/digital-edition/)
- Microwaves & RF (www.mwrf.com/members/magazine-digital-archive)
- Oil & Gas Engineering (www.oilandgaseng.com/magazine/)
- Plant Engineering (www.plantengineering.com/magazine/)
- Plant Services (www.plantservices.com/issue-archive/)
- Process Technology (www.processonline.com.au/magazine)
- Smart Industry (www.smartindustry.com/issues/)

Why study these resources? If you work, for example, in the Oil & Gas industry, you must know what the buzzwords mean in that industry. For example, do you know what the terms upstream, midstream, and downstream mean in that space[24]? If not, you'll be lost when partnering with the OT professionals working on a wireless IoT solutions project. These resources will teach you the language of the industry they represent, but all of those referenced here also have frequent articles related specifically to IoT, making them priceless resources in our profession. Most of the listed resources are free, except for the IEEE resources, and require only an email registration to access them.

The list is very long and more could be added, but you will find that a quick review of each one as it is released, taking no more than 5-10 minutes, will suffice and on frequent occasions an article will catch your attention that you want to read fully. That is exactly the use for keeping up with this rapid IoT space.

---

[24] In the Oil & Gas industry, upstream is a reference to the source of the resource (oil wells, etc.), midstream is a reference to the transmission of the resource (pipelines, hauling, etc.), and downstream is a reference to the processing of the resource into fuels in production companies.

Social Media

Another area that you can get involved in is the social media space. Among technologists, Twitter is very popular. Communities related to security, Wi-Fi, and IoT can be found there (as well as many others in the tech space). You can find other professionals with the same interests and begin by following them. Over time, you'll see the others involved in the community and you will eventually see the larger picture of the community that exists there. Most such communities are very helpful in answering questions and providing valuable feedback. So, get involved.

An additional way to get involved is to give to the community. Write useful blog posts and share them on social media. Create content that can be shared on LinkedIn, Facebook, Twitter, and even YouTube. Get involved in sharing information and the community will be glad to participate.

Beyond the Exam: Tom's Top Five YouTube Channels

Finally, I want to share my personal favorite YouTube channels related to IoT. While I sadly came to know about most of these later in my learning and practical work processes, they are the top five that I recommend to others on a regular basis. I hope you find them useful as well:

CWNPTV: Does this one surprise you? Years of historical webinars, commentary, tips, tutorials, and other information related to wireless can be found here. (www.youtube.com/user/CWNPTV)

4.0 Solutions: The channel of Walker Reynolds providing excellent videos with explanations of industrial technologies old and new. (www.youtube.com/c/IntellicIntegration)

Institute for Manufacturing (IfM), University of Cambridge: This channel provides a wealth of knowledge related to digital manufacturing and IoT. (www.youtube.com/c/ifmcambridge)

MultiTech Systems: While not having as many videos as the other channels listed, they do offer some insights not found elsewhere with some good videos on LoRaWAN and IoT connectivity. (www.youtube.com/user/MultiTechSystems)

The Things Network: While focused primarily on LoRaWAN, a significant amount of great IoT knowledge is presented here on a regular basis. (www.youtube.com/c/TheThingsNetworkCommunity)

I could have provided a list as long as the preceding list of magazines, but I'm sure you can find others of value. The key to staying ahead in the knowledge of IoT is to explore, practice, and repeat.

-Tom

1.6: Chapter Summary

In this chapter, you began by exploring the current and likely future state of IoT. Then you learned the wireless IoT solution design process from a high-level viewpoint and covered detailed areas of the design phases: Define, Design, Deploy, and Validate and Optimize. You should also be familiar enough with the steps of project management so that you can effectively manage your project from start to finish within the time frame and budget given. Finally, you reviewed a plan for keeping up with the change in the IoT space using freely available online resources.

1.7: Review Questions

1. What role does the gateway typically play in a wireless IoT solution?

    a. Storing data for analysis

    b. Connecting the wireless IoT network to another network

    c. Compute processes for edge processing

    d. Monitoring actuators for control of attached machinery

2. Why might you perform pre-design in the Design phase to gather information about the target environment?

    a. Spectrum analysis

    b. Define configurations for devices

    c. Create a BOM

    d. Specify the configuration and capabilities of supporting services

3. What is one reason it is important to discover where end devices will be located in a wireless IoT solution?

    a. To ensure that all devices support mobility

    b. To enable the spectrum analysis features of the device

    c. To determine the encryption to use on the wireless link

    d. To determine how the device will be powered

4. True or False: All IoT solutions use the Internet.

    a. True

    b. False

5. A post-installation survey is part of which phase of the design process?

    a. Define

    b. Design

    c. Deploy

    d. Validate and Optimize

6. Which one of the following is an example of an IoT Data protocol?

    a. MQTT

    b. 6LoWPAN

    c. ISA100.11a

    d. LoRaWAN

7. Which of the following are standards bodies related to IoT? (Choose three.)

    a. IETF

    b. IEEE

    c. ISO/IEC

    d. Control Engineering

8. In order to have a properly functioning wireless IoT solution, there is no substitute for a _____.

    a. Consulting Firm

    b. In-House Designer

    c. Spectrum Analyzer

    d. Good Design

9. True or False: An IoT design without the Internet is likely to be more complicated because public cloud services cannot be used?

    a. True

    b. False

10. What three phases of the wireless IoT solution design process are most likely to involve recursion?

    a. Define, Design, and Deploy

    b. Define, Deploy, and Validate and Optimize

    c. Design, Deploy, and Validate and Optimize

    d. None of these

1.8 Review Answers

1. **The correct answer is B.** The gateway is the device to which the end devices connect wirelessly. In most cases, it either forwards the data across the attached IP network, if the device is transmitting higher layer IP packets or it converts the data to an IP packet payload to be forwarded to the proper destination. With some systems, it uses proprietary protocols to communicate with other network devices on behalf of the end device.

2. **The correct answer is A.** Spectrum analysis is used to locate interferers, either intentional or unintentional in the target space. It is also used to identify the optimal frequency band to use for the IoT solution and, therefore, the possible protocols that will work well in the space.

3. **The correct answer is D.** Knowing the location of end devices allows you to plan for power provisioning on the devices. If AC power is not available, another powering solution will be required. PoE is not a practical solution for powering wireless IoT end devices. It PoE is used, in most cases, the wired Ethernet connection can also be used for data transfer. In some rare scenarios, a PoE injector may be used to power the end device without having an Ethernet connection on the wire and then wireless IoT protocols may be used for communication.

4. **The correct answer is B.** Many solutions in the IoT category do not use the Internet. They simply provide network connectivity and some do not even use IPv4 or IPv6 at all.

5. **The correct answer is D.** The post-installation survey, also known as a validation survey, is performed during the Validate and Optimize phase. Here the wireless IoT solution is tested, adjusted, and tested again to ensure customer requirements are met. This phase comes after requirements are defined and the design is completed and deployed.

6. **The correct answer is A.** MQTT, AMQP, and DDS are examples of IoT Data protocols. LoRaWAN, 6LoWPAN, and ISA100.11a are wireless IoT protocols and do not define specific application data, which is transmitted in their payloads.

7. **The correct answers are A, B, and C.** The IEEE, IETF, and ISO/IEC all create standards related to IoT including IoT Data protocols, wireless IoT protocols, network and transport protocols, and requirements engineering standards.

8. **The correct answer is D.** While the other options may be beneficial in certain circumstances, a good design is the key to a well-functional wireless IoT solution.

9. **The correct answer is A.** The IoT without the Internet is more complicated because it means you cannot consume public cloud services, which handle all of the hardware and much of the software configuration and implementation procedures.

10. **The correct answer is A.** The Define, Design, and Deploy phases will likely involve some recursion, particularly in complex IoT solutions. The Validate and Optimize phase is mostly self-contained. You are validating that the solution meets the requirements and, if it does not, you optimize it so that it does. It is less common to go back to the Define phase and change requirements at this stage, though in rare cases, a design change may be required. Typically, such design changes have surfaced during deployment rather than after.

Chapter 2: Assessing an Existing IoT Solution

Objectives Covered

1.1 Evaluate an existing IoT implementation and understand its impact on a new wireless IoT deployment
1.2 Use appropriate tools to analyze existing IoT implementations
1.3 Gather system documentation for the existing IoT solution
1.4 Evaluate operational parameters
1.5 Document findings for use in the design of the new wireless IoT solution

Evaluating existing IoT solutions is required anytime new IoT solutions are implemented. Several reasons drive this reality. First, reuse is important in fast changing environments, like IoT, and knowing what's already there can prevent the unnecessary work of planning and performing installations of redundant services. Second, given that we are focused specifically on IoT solutions that use wireless for the link layer, the frequency bands used by the existing solutions will impact your decisions for the new solution. You may choose to use the same protocols and layer your solution on top of what is there, or you may choose to use different protocols, possibly in different frequency bands, to avoid interference. Finally, evaluating the existing IoT solutions provides insight into current operations and where they might be improved, bringing added value to the new solution under planning.

In this chapter, you will explore the process used to evaluate existing IoT solutions. You will learn about tools used to analyze the wireless medium and determine frequency bands in use as well as discovery tools and documentation that allow you to determine services existing on the network for the support of the IoT solutions. The information gathered will be documented and useful during the requirements engineering process that is to follow[25].

2.1: What is an Existing System?

When evaluating existing systems, the design professional should not limit the assessment to those systems that are called IoT. IoT uses technologies that are used by other systems and must integrate with technologies already in place. For these reasons, the design professional must discover both IoT and non-IoT systems of interest within the target space and organization. We will review non-IoT systems first and then explore IoT systems.

---

[25] Evaluating existing solutions is sometimes performed before any requirements engineering and sometimes after business and stakeholder requirements have been defined but before or during the time when technical requirements are produced. Regardless of when you perform the processes, they remain largely the same. We have placed the information here in the flow of the book to begin introducing technical tools used in the design process and to provide this preliminary information to aid in the understanding of concepts addressed in the requirements engineering chapters to follow.

Existing Non-IoT Systems

The first category of non-IoT systems that should be assessed are other wireless systems. These are wireless communications systems that do not fit into the category of IoT, such as IEEE 802.11 Wireless LANs, wireless phone systems, wireless video surveillance systems not used with IoT technologies, and systems that intentionally radiate microwaves for purposes other than communications, such as microwave ovens. These systems should be discovered in the space to ensure that your wireless IoT solution can work around them or in cooperation with them.

The second category is the systems in place that perform functions the new IoT solution may be enhancing or replacing. These include older wireless sensor networks (WSNs), Supervisory Control and Data Acquisition (SCADA) systems, Programmable Logic Controllers (PLCs), and Distributed Control Systems (DCS). These are systems used in industrial, manufacturing, oil & gas, and other environments used to monitor and control machinery and the environments in which it operates. When implementing an IoT solution to replace these systems, the organization is almost certainly going to desire the same functionality with enhancements. When coexisting with these systems, the organization is likely to desire that the IoT solution share data with those systems or retrieve data from those systems and integration will become a factor.

To learn about the features and capabilities of the second category of systems, the wireless IoT solutions designer should meet with key OT professionals[26] and users of the systems to understand how they are used, what they provide, and what enhancements may be desired. This task will be part of the requirements engineering process explained in Chapters 3 and 4, but the process of assessing existing systems reveals their presence and the need to consider them.

---

[26] If you are an IT professional, the OT professionals will prove priceless as you implement Industrial IoT solutions. If you are an OT professional, the IT professionals will prove priceless as you implement the same. The convergence of IT and OT is a process occurring quickly in this decade and is likely to continue. We saw much movement in this direction throughout 2020 and 2021 with the COVID-19 crisis and the need to quickly implement many solutions that required the historic skillsets of both groups. Given the "proof in the pudding" of these exercises and the rapid implementation of complex systems, organizations will learn from this and are likely to continue having these groups work more closely together.

The third and final category of non-IoT systems is the network support services and applications in use. The IoT solution you design will require specific network services and capabilities. Many of these will likely exist in the environment and can simply be used by your solution if they have sufficient remaining capacity. Additionally, if the organization uses a public cloud service, such as Microsoft Azure, AWS, or Google Cloud, it is useful to know this. You don't want to unnecessarily require an AWS cloud account for your IoT solution if a Google Cloud service will work equally well and they already have an account with that provider and are satisfied with the service. Of course, you should always verify with the organization that they plan to continue using the existing account and that they do not prefer to spread their cloud subscriptions among multiple providers. Some organizations intentionally use multiple providers to prevent "having all their eggs in one basket."

Existing IoT Systems

Existing IoT systems may be wired or wireless. Both systems may use similar higher layer services but have different lower layer protocols within the network stack. For example, WirelessHART was created to provide wireless lower layers for HART-based communications. HART is a wired protocol and WirelessHART is wireless, but they can both support the same upper layers. In the same way, an Ethernet IoT device may use IPv4 or IPv6, both of which can be used on various wireless lower layer protocols. The point is that existing IoT systems, though wired, may already implement some of the higher layer functions required by your new wireless IoT solution.

Some protocols are used with many different lower layer IoT protocols. These include the IoT Data protocols discussed in the preceding chapter, like MQTT and AMQP[27]. In the same way, IoT solutions can use the same databases used for other purposes, whether they be SQL databases (like Oracle, SQL Server, and MySQL) or NoSQL databases (like mongoDB, DynamoDB, and Cassandra).

Existing wireless IoT systems must also be considered from an RF-level perspective. If they operate in the same frequency band as your desired IoT solution protocols,

---

[27] IoT Data protocols that are popular today include MQTT, CoAP, AMQP, DDS, HTTP/HTTPS, and WebSocket. The point of these protocols is a standard method for communication of data payloads regardless of the underlying network. They are covered in more detail in Chapter 7.

interference between the systems is a concern[28]. Knowing the RF in use by these existing wireless IoT systems is essential.

In summary, an existing system of interest is any system that may provide resources for reuse within your wireless IoT solution or may cause problems at any network layer for your solution or may be impacted by your wireless IoT solution. We will discuss more about the purpose of the assessment of these systems in the next section on evaluating operational parameters and the importance of gathering documentation, using tools, and documenting the results of your analysis in the remainder of the chapter.

2.2: Evaluating Operational Parameters

Operational parameters are the specifications of the existing systems. These parameters define how the system operates and the functional capabilities provided by and for the system. This section will address operational parameters including frequencies used, wireless signal coverage, functionality, and potential impact on new deployments as each parameter is discussed.

Frequencies in Use

You don't care about what you don't care about. The frequencies used by your potential wireless IoT protocols are the concern. These two direct statements are intended to give pause. There is no need to perform a spectrum sweep (detection of active signals and energy) across every frequency band if you already know the protocols and frequency bands you will be using. You are concerned only about the bands in which you may select a wireless IoT protocol for implementation. However, when you have not yet chosen the target protocol, you may need to sweep the entire set of likely bands and identify activity in them. The most common bands used by wireless IoT protocols include the following:

---

[28] If you are performing the assessment before requirements engineering or even early in technical requirements engineering, you may not know the IoT protocols that will be used as yet. This condition does not pose a problem to identifying existing wireless protocols in use. In fact, know that an environment already used Zigbee, for example, may be a factor in selecting the protocols for your design. Using Zigbee, given the availability of existing coordinators, may be the best choice.

- 800 MHz
- 900 MHz
- 2.4 GHz
- 3.5 GHz
- 5 GHz
- 6 GHz

These are the bands most wireless IoT protocols operate in. The 5 GHz and 6 GHz bands are those used by Wi-Fi and a few proprietary wireless technologies, but little in the way of other potential wireless IoT protocols. The 3.5 GHz band is used for CBRS (Citizens Band Radio Service) and is used mostly for private LTE solutions in the IoT space. It provides for incumbents to continue operations and others to have a priority access license or to use general authorized access. The 2.4 GHz band is used by several potential wireless IoT protocols including:

- Wi-Fi
- Bluetooth Low Energy (BLE)
- 802.15.4 protocols[29]
  - 6LoWPAN
  - Thread
  - Zigbee
  - ISA100.11a
  - WirelessHART

In the 800 MHz and 900 MHz band, you have the long-range wireless IoT protocols like LoRaWAN[30] and Sigfox.

---

[29] Some 802.15.4 implementations may use the 900 MHz and 800 MHz bands, for example, Zigbee (older devices), Thread (in Australia and Europe), and a few others. The significant move is toward more 2.4 GHz 802.15.4-based protocol implementations. In fact, it is harder every month to locate Zigbee modules based on the lower bands. Of course, with Zigbee being seemingly abandoned with the rebranding of the Zigbee Alliance to the Connectivity Standards Alliance, its future is questionable. No, they didn't completely abandon Zigbee, but it has become a small percentage of the focus of the new alliance, so time will tell what happens with it.

[30] LoRaWAN also supports the 470-510 MHz band in China.

While not in the preceding list, some protocols may use the 433 MHz and 470 MHz bands, such as LoRaWAN in some regulatory domains and a limited number of 802.15.4 devices used in IoT.

In addition to the standards-based protocols discussed, several proprietary protocols are commonly used. For example, Silicon Labs sells proprietary wireless SoCs (System on a Chip) based on various processors (mostly ARM Cortex) that operate in ranges from 110 MHz to 2.4 GHz. Interestingly, many of these proprietary protocols seem to be very similar to 802.15.4 in the 2.4 GHz band. For example, the EFR32FG13 (Figure 2.1) 2.4 GHz radio supports Gaussian Frequency Shift Keying (GFSK) modulation at 125 kbps or O-QPSK DSSS modulation at 250 kbps, which are the same rates supported by both modulations in 2.4 GHz in the 802.15.4 standard. When you dig deeper into the documentation for this module, you will read the following:

> Due to the shorter preamble of 802.15.4 and BLE packets, RX diversity is not supported[31].

Figure 2.1: EFR32FG13 2.4 GHz and 868 MHz and 915 MHz Radio Boards[32]

---

[31] www.silabs.com/documents/public/data-sheets/efr32fg13-datasheet.pdf
[32] Source: www.silabs.com/wireless/proprietary/efr32fg13-series-1-sub-ghz-2-4-ghz-socs

And there you have it. The proprietary wireless SoC uses 802.15.4 for O-QPSK DSSS modulation. At least this vendor hints at it. Many clearly seem to use it (as seen with some tests using 802.15.4 capture devices and being able to successfully decode frames) but lack any suggestion of it whatsoever. The moral of the story is that many proprietary protocols are only truly proprietary in the upper layers, and they often use radios based on existing standards.

Regardless of the actual protocols used, it is essential that you know two things about the frequency bands used:

- How much of the band is in use?
- What is the utilization in those channels/bands?

From this, you can plan for your wireless IoT solution. If the channel or band is only utilized at ten to twenty percent, you may implement your solution in the same band with no conflicts, particularly if both protocols used "listen before you talk" mechanisms. If they use a time-slotted mechanism where devices communicate on a schedule without performing a carrier sensing process, interference may still occur, but with significantly small utilization and if both systems allow for retries, the solutions may still work well together. You must know the requirements of your system and the impact of the existing systems to make this decision.

Wireless Signal Coverage

When existing wireless transmitters are detected in the target area, in addition to discovering the frequencies and utilization, it is important to know the portions of the target area (physical space) covered by the signals. This can also be discovered with a spectrum analyzer, but it is not the ideal tool. A wireless survey tool that can monitor the signal detected would be ideal. Such tools may be expensive or unavailable for the protocol of choice, though certain tools may support a spectrum adapter that can at least capture RF energy levels as you move throughout the facility. To effectively use a spectrum analyzer for this purpose, it should have a high resolution and support the frequency bands in question. Ideally, a spectrum analyzer that supports analysis from approximately 300 MHz to 7 GHz would be used as this covers nearly all the frequency bands used by IoT protocols.

Functionality

The next operational parameter set is focused on functionality of the existing systems and the components that provide this functionality. These components may be required by the new wireless IoT solution as well and may be used to support the new system. First, we'll consider network servers and services used in the existing systems of relevance to a wireless IoT solution and then we'll consider the protocols implemented.

Network servers and services in use can assist in supporting a newly deployed wireless IoT solution. Common servers and services that may be in existence and used by the new solution include:

- **Database Servers:** These include bot SQL databases and NoSQL databases and other odd types of database systems. To determine if the database servers in place can be used by the new IoT solution, consult with the database administrator to discover current capacity levels utilized. If sufficient capacity remains, the database server may be used for the new system.
- **IoT Data Protocol Servers:** These include the IoT Data protocols referenced earlier and are the servers that support their operations. An existing IoT administrator or engineer may provide capacity utilization information on these servers to determine their possible use for the new system.
- **Authentication Servers or Appliances:** These range from large-scale authentication systems, like Microsoft domains to AAA authentication systems used mostly for network devices. The security or network administrator is likely to know the available capacity for these services.
- **Time Services/Servers:** The most common service used for time synchronization on networks is the Network Time Protocol (NTP). They typically fall under the management control of network administrators.
- **SMTP Servers:** Email servers are often used to send notifications and alerts by IoT systems and can typically handle large loads that are queued. An existing server is likely to be able to handle the additional needs of an IoT solution but verify this with the email administrator.
- **DHCP/DHCPv6 Servers:** If an IPv6-based IoT solution is planned, DHCPv6 servers may be required. It will depend on the architecture chosen. You may choose to implement a solution where the border routers assign IPv6 addresses to the IoT nodes, and the nodes never communicate directly with the rest of the

network but use the border router as an effective proxy. Depending on your plan, the network administrators can assure that sufficient IPv4 or IPv6 addresses are available in the DHCP pools within the network.

- **Cloud Services:** If cloud services are to be used for the IoT solution, either specific cloud IoT services commonly available in public cloud providers or simply databases and services in the cloud, two factors should be evaluated. First, the existing cloud service account may be used to fulfill the requirements, if desired by the organization. Second, the Internet connection must provide sufficient capacity to handle existing workloads and new workloads introduced by the IoT solution.

When considering the protocols implemented, there is some overlap with the preceding list. For a typical wireless IoT solution, you must first know the wireless protocols in use within the target space and the frequency bands and channels used. Second, you must know the network protocols in use within the larger scale of the overall network. An existing protocol architecture can be used for the transmission of data from IoT devices to the intended destination (either on-premises or in the cloud). The protocols in reference here are protocols like IPv4, IPv6, Ethernet (with QoS), routing protocols, etc. They are the protocols that may support the operations of the larger network to which the IoT solution will connect.

The final area to consider with implemented protocols is the IoT Data protocols referenced previously. If an architecture is in place supporting MQTT, DDS, CoAP, HTTPS, AMQP, or WebSocket, the IoT solution may use the existing implementations. Keep in mind that CoAP and HTTPS are request/response protocols and "existing systems" would simply be CoAP devices with which the IoT devices may communicate or web servers (HTTPS)/REST API servers.

2.3: Gathering Documentation

One source of the information discussed in section 2.2 is existing documentation. Network diagrams may reveal sufficient information about the services and servers running on the network or at least the structure of the Layer 1-Layer3 network services. System architecture documentation may also provide insightful information. It will often list the servers implemented, protocols used, and various interface connections between systems. An additional source is requirements specification sets for existing systems. If the existing systems were implemented with proper requirements engineering and then validated to meet those requirements, the requirements specification sets will reveal the services and capabilities in place within the organization. A post deployment validation and test report would be even better as it would list the requirements implemented in the solution as opposed to those that may have been referenced in the initial requirements specification but, for one reason or another, were not implemented.

In addition to knowing "what" is there, important quality requirements can be discovered by viewing security policies[33]. Security typically falls under the quality requirement set, as discussed in the next two chapters. The security policies will reveal security requirements for any new system. The policies, if properly documented, will be generic enough such that they can apply to multiple solutions. Some policies may specify things like the use of AES-128 as a minimum encryption level. If you can provide several different methods to accomplish this, you have achieved compliance with the generic policy. For example, you can pre-stage end devices with key materials or key materials can be generated during onboarding. If both methods result in AES-128 encryption, the policy has been met.

2.4: Tools for Analyzing Existing Systems

In preceding sections, you discovered what you need to assess. In this section, we will discuss how you can assess it. For the purposes of this chapter, we will constrain the tools to protocol analyzers, spectrum analyzers, and network diagrams. In all three categories, significant differences exist among the tools and knowing how to select the right tool is important. Each tool will be covered from the perspective of common features and the use of the tool for the assessment of existing systems.

Protocol Analyzers (Wired and Wireless)

Protocol analyzers fall into two general categories: frame analysis and information reporting. Frame analysis analyzers allow you to see the specific individual frames by

---

[33] Several quality requirements are likely to be identified during requirements engineering. The wireless IoT solution designer should be familiar enough with the entire IoT architecture to define the quality requirements for a given solution.

capturing them from the network and may be wired, wireless, or both. Information reporting analyzers provide you with information about the protocols on the network but do not reveal the actual frames in use. The former is what we typically call a protocol analyzer, and the latter is what we call a scanner, monitor, or troubleshooting tool. However, many tools are sold for IoT protocols that are called protocol analyzers but lack the ability to see the individual frames. Always verify what the tool can do with vendor documentation and feature lists.

Without question, Wireshark is the most well-known protocol analyzer. In part because it's free, but also because it can decode many different network frames and packet types. Specific protocol analyzers are available for different protocols[34]. In some cases, you may have to build a solution to meet your needs (when budget is limited) and in other cases you can purchase off-the-shelf solution ready to do analysis. When building a protocol analysis solution, you typically start by identifying a dongle or board that can capture the target protocol and then ensure that software is available allowing it to capture the frames and store them in a format compatible with protocol analysis software such as Wireshark. Dongles are typically USB-based and boards may be single-board computers that support add-on modules, such as RaspberryPI devices, or they may be maker boards that have built-in chipsets for a given protocol.

An example of building a solution is the use of the Texas Instruments CC26x2R LaunchPad device (Figure 2.2). This device supports 802.15.4 in the 2.4 GHz band and incorporated a printed circuit board (PCB) antenna. At the time of writing, the device is $39 US. This device can be coupled with the Ubiqua Protocol Analyzer (priced at $65 US per month at the time of writing) to capture and decode 802.15.4 communications, including Zigbee, Thread, and 6LoWPAN. Figure 2.3 shows an example of the packet decode comparison tool within this analyzer.

---

[34] In the IoT space, as many different protocols exist, multiple hardware and software solutions may be required to enable protocol analysis for the many different protocols. It is not uncommon to require one kit for 2.4 GHz analysis (being careful to select one that can do BLE and 802.15.4-based systems) and another kit for sub-1 GHz. With careful planning, a single kit may be able to do analysis for all protocols but will likely require swapping modules with something like a Raspberry PI.

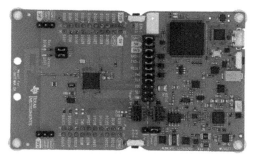

Figure 2.2: TI CC26x2R LaunchPad 2.4 GHz Device with Onboard PCB Antenna

Figure 2.3: Comparing Packet Decodes in Ubiqua[35]

An example of an off-the-shelf solution is the Sewio Open Sniffer shown in Figure 2.4. This analysis hardware supports integration with Wireshark (and can also work with Ubiqua Protocol Analyzer) and capture of 802.15.4 protocols with dissectors (decoders) for Zigbee and 6LoWPAN. It also supports sniffing on the sub-1 GHz bands.

---

[35] Source: ubiqua.io

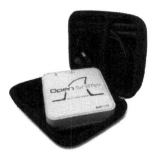

Figure 2.4: Open Sniffer from Sewio

Once configured and connected to the Ethernet network port of a laptop, it can capture 802.15.4 frames in the area where it is located. Professionals access it through a web-based interface to initially configure it and then point Wireshark to the Ethernet interface on which it is connected for capture. Finally, filter to see only WPAN traffic and you will be presented with 802.15.4 captures in the Wireshark analyzer as shown in Figure 2.5. Open Sniffer captures the 802.15.4 frames on the radio interface and forwards them to the laptop computer. Wireshark is performing the decode operations. At the time of writing, Open Sniffer was priced at about $380 US.

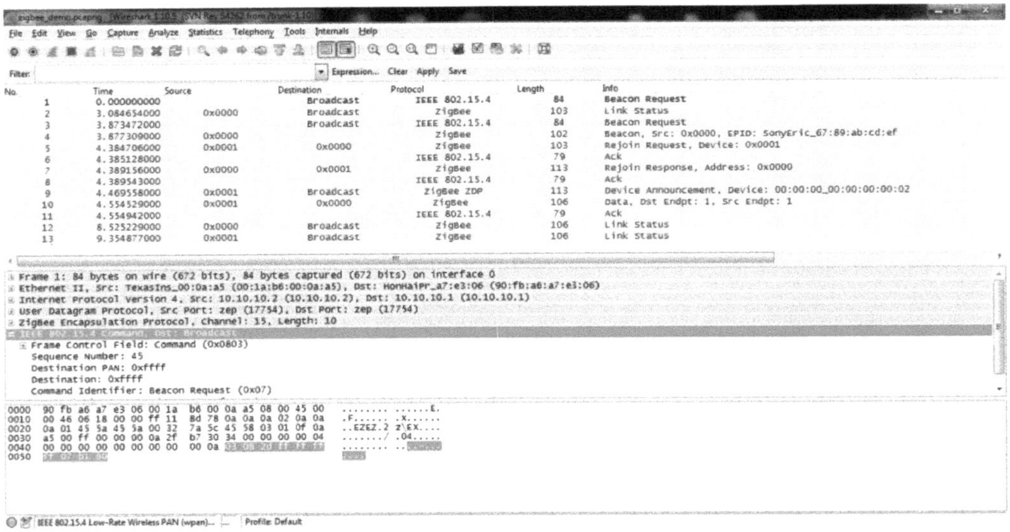

Figure 2.5: 802.15.4 in Wireshark

6LoWPAN can be used with the Routing Protocol for Low-Power and Lossy Networks (RPL)[36], which uses a Directed Acyclic Graph (DAG) and specifically a Direction-Oriented DAG (DODAG). *Acyclic* indicates that it is non-cyclic or doesn't contain cycles or loops. All connections in the graph point in directions that never close the loop. The concept comes from graph theory where a *directed graph* is a set of *vertices* and *directed edges* (also called *arcs*), which connect a pair of vertices. When a directed path starts and ends with the same vertices, it is a directed cycle. When no path is a directed cycle, it is acyclic. When direction-oriented, it indicates that the graph is rooted in a single destination, which, itself, has no outgoing directed edges. To simplify it for 6LoWPAN, the vertices are the nodes, and the directed edges are the connections between them. The root is typically the border router that connects the 6LoWPAN network to other networks.

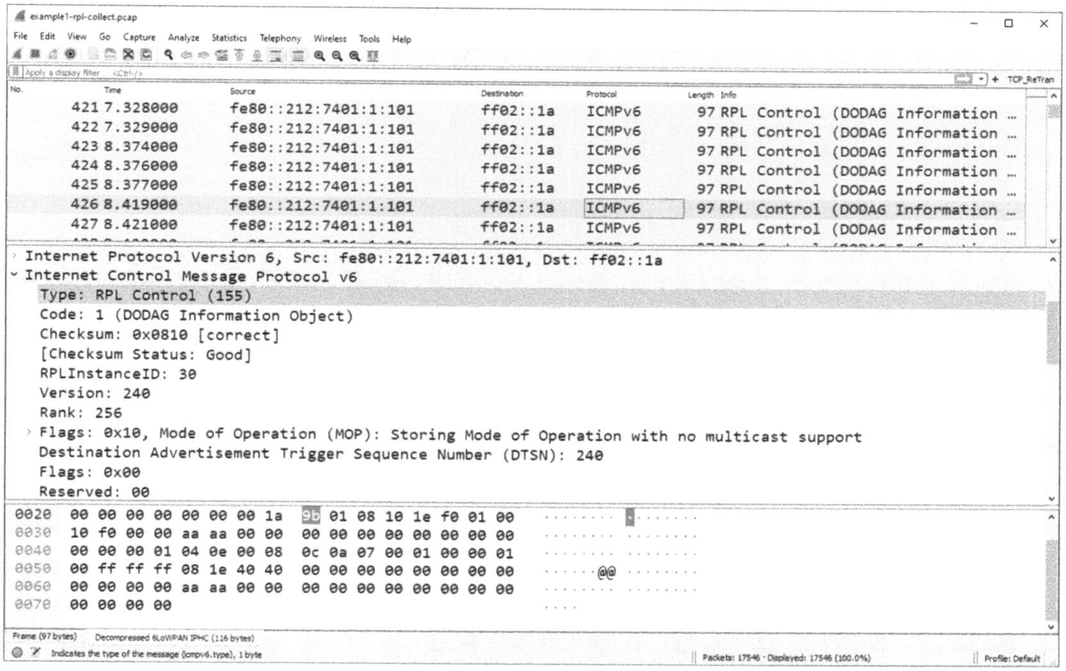

Figure 2.6: DODAG RPL Exchange Shown in Wireshark

---

[36] RPL is defined in IETF RFC 6550, RPL: IPv6 Routing Protocol for Low-Power and Lossy Networks

Why all this information about how 6LoWPAN builds a routing structure? First, it's useful to know this if you must deploy a 6LoWPAN solution. Second, it's very useful to know that this information can be seen traversing the network in Wireshark. Wireshark has decoders built in for 6LoWPAN, Zigbee and a few other IoT protocols right after installation. Figure 2.6 shows a decode of an RPL packet used to build the DODAG tree and you can see that the packet contains a DIO (DODAG Information Object), which is used in the development of the graph.

How is this used in assessment of existing systems? The answer lies in understanding that many protocols use 802.15.4; however, when you see DODAG messages and then also see IPv6 and even explicit 6LoWPAN messages, you know that the network is indeed using 6LoWPAN. So, building a frame capture tool that can capture 802.15.4 frames in 2.4 GHz and then opening the PCAP in Wireshark will reveal whether the higher layers are 6LoWPAN, Zigbee, ISA100.11a, WirelessHART, and even proprietary protocols in some cases.

Using a protocol analyzer on the wired network can reveal services and capabilities of that network as well. However, due to the nature of wired communications today in switching infrastructures, you would likely need the assistance of a network administrator to enable forwarding of Ethernet frames to the port on which you're monitoring.

Of course, always, always gain permission before using any monitoring tools within an organization, including protocol and spectrum monitoring. One might think that spectrum monitoring wouldn't reveal sensitive information, but it can certainly reveal characteristic traits of wireless communications that would indicate the type of wireless protocol in use. Again, always gain permission first.

Protocol analyzers are the software tools used to analyze the captures from a network. The capture can be performed by the analyzer or by another tool and then loaded into the analyzer. The job of the analyzer is to parse the captured data, decode it for easier interpretation, and present the information. Some analyzers have more features than others, but most will offer filtering to select only the frames or packets desired, exporting to a format for use in other analyzers, importing from common capture formats (like PCAP and PCAPNG), and searching to locate specific communications.

More advanced features may also be available. These include:

- **Statistics:** The analyzers will often provide several types of statistics for summary analysis. These can include counts of end devices, conversations (communications between specific devices), packet lengths, throughput, destination, and source addresses and more.
- **Graphs:** Many analyzers will also present information in graphs that can be used to summarize the previously mentioned statistics. Such graphs can be useful when building documentation of your existing systems assessment.
- **Protocol-Specific Experts:** If the analyzer is targeted at a specific protocol, like 6LoWPAN, Wi-Fi, or LoRaWAN, it may provide expert analysis tools that investigate and report on captured information that impacts performance, security, or general communications within the context of that protocol.

When selecting a protocol analysis solution for your projects, it must meet the following requirements:

- Be able to capture the protocol desired
- Be able to tune to the frequency band and channel desired
- Be able to decode the protocol desired

Spectrum Analyzers

Spectrum analyzers capture and display RF energy. A spectrum analyzer that works to monitor RF energy "in the air" is needed to evaluate existing wireless activity in the target space. Desktop spectrum analyzers that show RF signals entering them through a cable do not work for this purpose. The desired analyzer will be portable and support the frequency ranges you wish to analyze. For IoT protocols, these ranges are typically from around 300-400 MHz to 7-7.5 GHz, if you wish to be able to capture spectrum activity from all the popular wireless IoT protocols.

The spectrum analyzer will capture RF signals detected through the attached antenna and display the characteristics of those signals in time-based views (waterfall or spectrographs) and frequency-based views (Fast Fourier Transform (FFT)). Over-the-air capture devices do not display time-based views that show the actual waveform (like a desktop analyzer). Instead, the time-based views show the power over time for each frequency. This view is the spectrograph or swept spectrogram or waterfall type view.

Figure 2.7 shows the inexpensive handheld spectrum analyzer, RF Explorer, monitoring a portion of the 900 MHz band and showing a signal at approximately 965 MHz. In Figure 2.8, we have the same RF Explorer used with the Windows application through a USB interface and you can see signal activity or RF radiation around 870 MHz.

Figure 2.7: RF Explorer Scanning 900 MHz

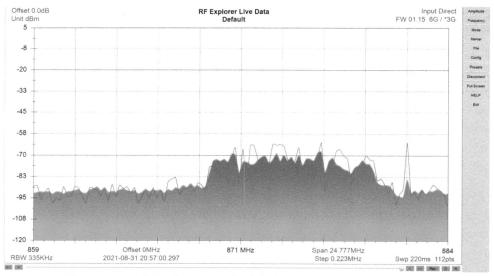

Figure 2.8: RF Explorer used with the Windows Application

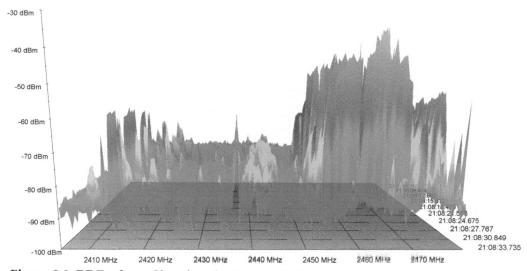
Figure 2.9: RF Explorer Showing the Waterfall View with RF Over Time in 2.4 GHz

Network Diagrams

Network diagrams, provided by the organization seeking a new wireless IoT solution, will provide valuable information. Varying levels of detail will be provided in such diagrams. In some cases, they present only an architectural view with no technical details of configurations. In other cases, they provide exceptional detail. Either way, the information they provide will be valuable though those with greater detail provide the greatest value.

Figure 2.10 provides an example of a low detail architectural diagram. In this case, the details are not provided. However, it does reveal important information assuming it is implemented as presented. First, we can learn that firewalls exist between the Internet and a demilitarized zone (DMZ)[37] and between the DMZ and the internal network. This could have implications if we are planning a cloud based IoT solution. We must ensure that the firewalls open the appropriate ports for the IoT communications.

---

[37] A DMZ is a perimeter network that typically sits between the internal LAN and the Internet. Servers that must be accessed by the Internet may exist in the DMZ or proxies may be implemented there to allow access to internal network servers.

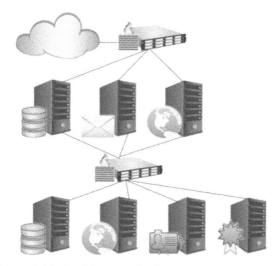

Figure 2.10: Architectural View Network Diagram

Second, we see that the network has two database servers. One in the DMZ and one on the internal network. This could indicate the availability of on-premises databases for our use. Finally, an email server is present as well as two web servers, an identity server, and a certificate server. If nothing else, even a simple architectural diagram like this informs me that resources may be available for reuse.

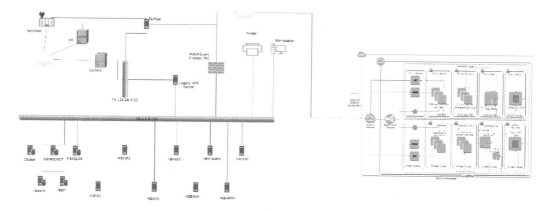

Figure 2.11: Detailed Network Diagram

In other, more pleasing situations for the IoT solution designer, the organization may provide a network diagram like that in Figure 2.11. Here we have details about the internal network and the AWS cloud services they use. The more detailed diagrams you can get, the better.

If the organization lacks detailed diagrams, part of the assessment should be querying the appropriate individuals to determine available switch ports and possibly Power over Ethernet (PoE) budgets available that might be used to power gateways in the wireless IoT network.

2.5: Documenting Assessment Results

Your final task is to document the results of your analysis. You have basically performed a true survey at this point. A survey or site survey is an analysis of a target location performed to determine the systems in that location that can be reused for your solution and that may conflict with some potential solutions and to determine the RF activity in the space. For a wireless IoT solution designer, a survey is not limited to RF, though the RF survey (which could be performed without other tasks in some scenarios) is part of the larger site survey[38]. The designer must survey the RF in the environment and how new RF signals will propagate through the environment most certainly, but with the demands of complexity for IoT solutions, the survey must accomplish more than simple RF analysis. It must meet the truest sense of the historic term, *survey*, to *inspect closely for the purpose of developing a map or a plan*. That is exactly what you're doing in an assessment of existing IoT solution and the existing environment: inspecting it closely for the

---

[38] It surprises many wireless engineers who have not worked in other IT disciplines to discover that the phrase *site survey* is not limited to RF activity. For example, Microsoft Active Directory engineers often use a site survey to discover the uses of the planned directory service. In the architecture world, a site survey is simply discovering and documenting the dimensions in a facility or on a campus. In hazardous waste and emergency response units, a site survey is performed when a violation or problem has occurred to assess the situation, create a plan to remedy it, and create a plan to prevent it in the future. An RF survey in the context of wireless IoT means analyzing the RF activity in an area and determining how planned RF will propagate in that area. A survey (whether called a site survey or simply a survey) for a wireless IoT solution includes this and much more.

purpose of developing a plan and a map (though we will call it a network diagram and a heat map, but the language works).

The survey results should include documentation of the following (limited, of course, to what you were permitted to perform):

- Spectrum analysis report, with screen captures of important RF activity.
- Protocol analysis report, with documentation of discovered protocols.
- Network diagrams provided by the organization.
- Discovered protocols, services, and servers that may be useful for the IoT solution design.
    - DHCP
    - NTP
    - IoT Data protocol servers/services
    - Database systems
    - Authentication servers
    - Email servers
    - Internet connectivity
- Available switch ports and PoE power for powering gateways when required.
- Photographs of important physical items and space constraints that may have been discovered.
- Additional information unique to the project that will be a benefit, constraint, or problem in relation to the new wireless IoT solution.

You may choose to document switches and routers available in a simple spreadsheet or to create your own rack diagrams if the organization cannot provide them. Figure 2.12 illustrates such a diagram. Tools like Visio and EdrawMax make the process of creating such diagrams simple and they prove extremely valuable later in the design process.

With the information gathered, you can best design a solution. As stated previously, the assessment may be performed before full requirements engineering or during requirements engineering or during the design phase, but it is an essential systems engineering process for wireless IoT solutions. The topic has been addressed here to begin introducing you to concepts and tools related to wireless IoT solution design and to address the importance of gathering this information to best determine requirements and design parameters.

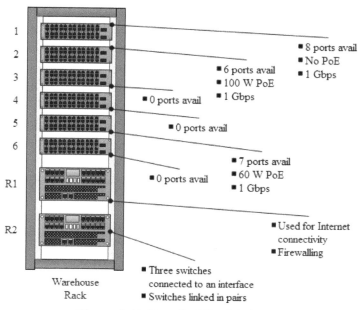

Figure 2.12: Detailed Rack Diagram

Beyond the Exam: A Brief History of Systems Engineering[39]

Like many things in technology, the *systems engineering* phrase goes back to Bell Telephone Laboratories in the late 1940s. The concepts of systems engineering go back further into the early 1900s within Bell. The concept of a *systems approach* was used within RCA in the development of the electronically scanned, black and white TV. The first use of the phrase *systems engineering* seems to be a memo within Bell in the summer of 1948 and by 1951 it was defined as a unique function in the organization's structure. The first course in systems engineering was offered at MIT in 1950 and the first book was written in 1957 by Goode and Machol. The book was *Systems Engineering – An Introduction to the Design of Large-Scale Systems*. As you begin to evaluate systems engineering and wireless IoT solution design, remember, you are not alone. Many have gone before, and answers are out there.

-Tom

---

[39] Dennis M. Buede. *The Engineering Design of Systems: Models and Methods, 2nd Edition.* 2009.

2.6: Chapter Summary

In this chapter you explored topics related to the CWIDP exam knowledge domain, *Assess an Existing IoT Solution* and went one step further to assess the environment in general with a full survey of technologies available for reuse in your IoT solution design. You also began learning about tools available for use in the analysis of existing wireless IoT solutions and other IoT solutions as well. You discovered operational parameters of the existing solutions and learned the importance of documenting the result of your assessment. In the next chapter, you'll begin using this information and other information you will gather to perform requirements engineering based on standard defined procedures.

2.7: Review Questions

1. Why might you choose to analyze existing wired systems as well as wireless systems during an assessment?
    a. The wired network may generate enough EMF to cause problems with your wireless IoT solution.
    b. The wired systems may implement technologies that can be reused.
    c. The wired systems may prevent implementation of wireless solutions because gateways are not typically compatible with wired systems.
    d. None of these.

2. If you discover existing wireless frequencies in use within an assessment area, what should you do?
    a. Analyze them to determine signal strength, coverage areas, and utilization levels.
    b. Ignore them and trust in the contention methods used in wireless protocols.
    c. Develop a plan to shield your gateways and end devices from these signals with wire mesh materials.
    d. Determine the security they use to ensure that the new wireless IoT solution will not inadvertently decrypt their payloads.

3. True or False: If an existing wireless system is at low utilization levels, it is possible that a new system can function well in the same frequency band or channel.
    a. True
    b. False

4. What is an example of an existing network service that would be useful specifically to 6LoWPAN wireless IoT solutions but not to many others?
    a. Instant messaging
    b. DHCP
    c. NTP
    d. Authentication services

5. True or False: IoT Data protocols include 802.11, 802.15.4 and 802.15.6.
    a. True
    b. False

6. What might you purchase to build your own wireless IoT protocol analyzer for 802.15.4 networks?
    a. A board or module compatible with the target wireless protocol and an analyzer that can capture the packets or frames.
    b. An 802.11 USB adapter because they are compatible with 802.15.4.
    c. An Ethernet adapter to capture the 802.15.4 frames on the wired side of the network.
    d. A US Robotic modem with an 802.15.4 chipset.

7. What is essential when selecting a spectrum analyzer for a wireless IoT protocol?
    a. That it can capture RF energy in the band used by the protocol
    b. That it can capture frames generated by the protocol
    c. That it can capture RF energy in all bands
    d. That it can integrate with a protocol analyzer

8. True or False: A network diagram can be useful even if it lacks detailed data.
    a. True
    b. False

9. True or False: Documenting the results of the assessment is an optional step because a small amount of information is gathered during the assessment of an existing IoT solution.
    a. True
    b. False

10. What document provides information related to quality requirements?
    a. Switch port configuration VLAN assignment details
    b. Security policy details
    c. Cloud service user account details
    d. Wall material specifications

2.8: Review Answers

1. The correct answer is B. The wired system will often implement technologies that can be reused for the wireless IoT solution reducing costs and time during implementation.
2. The correct answer is A. You cannot ignore detected wireless signals. Instead, you should evaluate them to determine the impact they will have on the new wireless system and to ensure coexistence.
3. The correct answer is A. When an existing wireless system has low utilization levels, it is possible, though not certain, that the new wireless IoT solution can coexist without problems.
4. The correct answer is B. DHCP, specifically DHCPv6, may be used by 6LoWPAN networks for IPv6 address and configuration distribution.
5. The correct answer is B. IoT Data protocols include MQTT, CoAP, DDS, AMQP, HTTPS, and WebSocket, among others. Wireless IoT protocols are not considered IoT Data protocols.
6. The correct answer is A. To build your own protocol analyzer, you will need a board, like a maker board, or a module that includes the chipset/radio for the target IoT protocol and software that can use the board or module to capture and analyzer the data.
7. The correct answer is A. The spectrum analyzer must be able to tune to and capture RF energy data in the frequencies used by the wireless IoT protocol.
8. The correct answer is A. A network diagram, no matter how basic, often provides insightful information that, at the least, can prompt questions you can ask of the organization to gain more information.
9. The correct answer is B. Documenting the results is essential because, while we've covered the basic process in a couple dozen pages, the results of the assessment could be hundreds of pages of information.
10. The correct answer is B. Security is a quality requirement. The other listed items are technical requirements or constraints or simply information.

Chapter 3: Gathering Business Requirements and Constraints

Objectives Covered

2.1 Gather business requirements and constraints

With the rapid evolution in the IoT space it can be difficult to match the best product, technology, or service with the needs of your company or customers. No single solution can meet the needs of every customer. In 2020 IoT-Analytics.com published a report showing that Manufacturing/Industrial is the largest IoT segment (when excluding the consumer market) followed by Transportation/Mobility, Energy, and Retail.

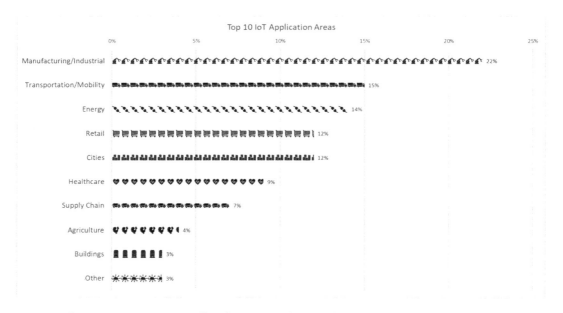

Figure 3.1: Top 10 Application Areas for IoT in 2020 (IoT-Analytics.com)

This IoT growth consists of many components that make it all possible. Powerful, but low-power-consuming, processing devices, standardized protocols at the network through to the application layers, cloud solutions, and many other technologies are allowing for this rapid growth. When preparing to create requirements for your project you will need to be well-versed in many of these areas. Since the job of technology is to enable the business, we will focus on the business requirements and stakeholder requirements before discussing how the technical requirements will enable the business requirements themselves. Requirements engineering is an essential part of the Define phase of wireless IoT solution design and may overlap with the design phase as recursion may be necessary.

Gathering business requirements and constraints for wireless IoT networks is similar to requirements covered in other types of networks, except for the fact that considerations above Layer 3 become more important to the design process than some other network types. This reality is, in part, due to the complexity of wireless IoT solutions and IoT solutions in general and, in part, due to the immaturity of the industry. Early on in wired networks, we were less aware of upper layer impacts on wired network needs. The same was true for 802.11 wireless LANs (WLANs). We are in that time window with IoT as well but are discovering best practices and ideal configurations quickly. With wired networks and WLAN networks, most user application types are well known and generalized: email, web browsing, voice, video, database access, etc. With IoT networks, the application types vary along the scale of payload sizes from less than a byte to several kilobytes, along the scale of delivery requirements from less than 1% packet loss to less than 20% (or more) packet loss, etc. As we continue to learn the application types and discover generalizations of them, complexity will continue during these early years of large-scale IoT deployments[40].

As a CWIDP you are expected to have a professional level understanding of the design of wireless IoT networks. Business requirements focus on "What do we need to do?" while technical requirements are "What are the functions, processes, and qualities required to deliver the business requirements?". In practice these tasks may be completed by the same individual or spread out over multiple people; however, the core principles outlined here for these functional areas are essential to ensure a successful design and deployment of an IoT network.

Here are a few questions that will need to be answered to ensure a successful design with your wireless IoT project (and this list is just a starting point):

- What are you trying to accomplish?
- What is the business purpose?
- What is the justification for the project?

---

[40] It's one thing to implement IoT in smart home environments, where a few dozen IoT devices are deployed. It's quite another things to implement IoT in smart factories, smart buildings, and other scenarios where thousands of IoT devices are deployed, and often in a space only eight to ten times larger than the smart home. When you have 20,000 devices deployed in 18,000 square meters of space, the density is significantly greater than when you have 50 devices deployed in 200 square meters of space. Some networks are planned or implemented that will have or have closer to 100,000 devices in 18,000 square meters.

- When does the project need to be completed?
- How much is the project going to cost or how much money is allocated for the project?
- Who will utilize the data generated from the IoT network?
- From where will the data be collected (physical location) and to where must it be transferred (data destination)?
- How often does data need to be updated?
- What areas need wireless coverage to serve IoT devices?

In this chapter we will define a common structure for requirements, take a deeper look at each of these questions, and go over what is commonly known as requirements engineering.

The process of creating requirements may require multiple iterations and the requirements may evolve over the time of the project. The level of effort, time, and detail that goes into gathering requirements will depend on how critical the wireless IoT system is that you are designing (Figure 3.2)[41] and how large it is. Some of these tasks may be skipped if there is a predetermined off-the-shelf solution to be installed. If you are designing an IoT system as part of a business, you will want to ensure you are considering the needs of every stakeholder (the individuals or organizations having a right, share, claim or interest in a system or in its possession of characteristics that meet their needs and expectations[42]).

Before we start to ask and gain answers to the questions above, let's focus on how to create requirements that are specific, measurable, solve the business problems, and that are based on involvement of the right stakeholders. This will require an understanding of the overall requirements engineering process and the specifics of writing individual

---

[41] Every wireless IoT solution design should have a requirements engineering phase, which we call Define in our model. It is often challenging to gain approval for the required time and budget to perform requirements analysis. It can help to ask questions like, "How will we know that the IoT solution meets your needs?" and "What is the ideal result of this project?" If you can get the stakeholders to begin thinking in this way, they are likely to be far more willing to move through the requirements engineering process. You are making it about assisting them in achieving their goals rather than just having meetings. The goal is to give them what they want, but you must discover what that is.

[42] IEEE 29148-2018 clause 3.1.28

requirement statements and complete requirement sets for a system or component of a system.

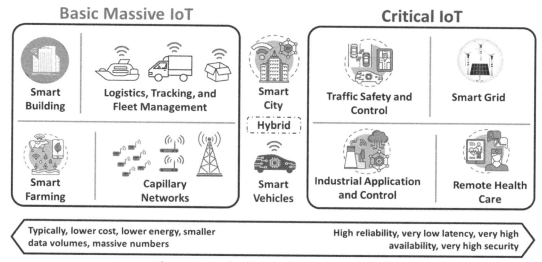

Figure 3.2: Massive IoT vs. Critical IoT[43]

3.1: Requirements Engineering Framework

In this section we will focus on what requirement are, how to create them, and form them properly. For the purposes of CWIDP, we will utilize *IEEE 29148-2018 Systems and Software Engineering - Life Cycle Processes - Requirements Engineering*[44] as our primary framework for both business and technical requirements. Throughout this section we will cite information from the IEEE standard and use it to guide examples of requirements. This standard and the supporting standards have been developed over a 49-year period and represent thousands of hours of collaboration, tens of thousands of hours of project experience, and many hundreds of projects among the professionals working on the committees over the years. A matured standard is far better than a single person's

---

[43] Enhanced from Infovista Whitepaper, *Industry 4.0 Connectivity Enabled Automation, Addressing the Challenges of Cellular IoT*
[44] We will refer to this standard as simply 29148-2018 throughout most of the rest of this material for simplicity.

experience and this is the primary reason that CWNP has chosen to standardize on the 29148-2018 standard for requirements engineering across all certifications[45].

The standard defines three requirements engineering processes[46]:

1. **Business or Mission Analysis (expanded in 29148 and outlined in ISO/IEC/IEEE 15288:2015):** The purpose of the Business or Mission Analysis process is to define the business or mission problem or opportunity, characterize the solution space (environment), and determine potential solution class(es) that could address a problem or take advantage of an opportunity[47]. Outcomes include:
   a. *The problem or opportunity space[48] is defined.* This definition may include political, economic, social, technological, environmental, and legal aspects (PESTEL). The problem or opportunity is clearly defined within the space.
   b. *The solution space is characterized.* The environment within which the solution will be deployed is characterized including the definition of primary stakeholders (both internal and external), the target operating environment (including known security threats in such environments, hazards, and even discovery of existing system), and the identification of candidate alternative solution classes.

---

[45] An additional benefit is that you know what you are being tested against. If you are preparing for the CWIDP, CWDP, or CWIIP exams, all requirements engineering and analysis knowledge comes from IEEE 29148-2018, IEEE 15288-2015 (System Life Cycle Processes), or IEEE 15289-2019 (Content of Life-Cycle Information Items (Documentation)). Requirements engineering is a soft skill driven by technical proficiency and the standards provide a solid foundation of knowledge for testing these skills and applying them in the real-world.

[46] IEEE 29148-2018 clause 6.1

[47] IEEE 29148-2018 clause 6.2.1

[48] The phrases *problem space*, *opportunity space*, and *solution space* come from the domains of marketing and product development. The *problem space* or *opportunity space* is where no product or solution exists. It is where unfulfilled needs exist, and a solution is needed. The *solution space* is where the system or product is coming into existence. It includes prototypes, proof-of-concept (PoC), and the actual system. During requirements engineering, you are working in the problem space. Early in design, you are still in the problem space, but once you start building the solution, including wireless design within a design application, you have entered the solution space.

    c. *The preferred candidate alternative solution class(es)[49] are selected.* This is achieved by assessing each candidate alternative solution class based on the defined criteria when characterizing the solution space. The assessment may include expert feedback, simulation and modeling, and other procedures. After assessment, the preferred solution class or classes have been selected.

    d. *Traceability of business or mission problems and opportunities and the preferred alternative solution classes is established.* The establishment of traceability in this early stage is essential so that stakeholder requirements can be linked back to the business or mission requirements and eventually the system requirements can be traced back as well. A requirements management tool may be used and identifiers[50] may be established, that will be used in the ensuing phases.

2. **Stakeholder Needs and Requirements Definition (expanded in 29148 and outlined in ISO/IEC/IEEE 15288:2015):** The purpose of this process is to define the stakeholder requirements for a system that can provide the capabilities needed by users and other stakeholders in a defined environment[51]. Outcomes include:

    a. *Stakeholders of the system are identified.* Many stakeholders were identified in process one; however, this process should begin by exploring potential new stakeholders as well.

    b. *Required characteristics and context of use of capabilities are defined.* The context of use includes the characteristics of the users, tasks and

---

[49] Solution classes may be defined in varying ways. One common method is to use the three classes of: operational change, system upgrade, and new system development. That is, a solution class may require only that people do work differently while another may require upgraded or new technologies.

[50] For example, if a business requirement (need) is identified such that an IoT solution shall be implemented to reduce accidents in a manufacturing environment by 15 percent, it may be stated as: *1.0: The IoT solution shall provide monitoring and control that results in a reduction of accidents resulting in human injury by 15 percent within six months of implementation.* The identifier (ID) is 1.0 at the top level (business level). Then, the stakeholder needs might be identified as 1.1, 1.2, … 1.*n*, where *n* is the final stakeholder requirement under the business requirement ID of 1.0. Next, the system requirements can be identified as 1.1.1, 1.1.2, … 1.1.*n* (under stakeholder requirement 1.1) and 1.2.1, 1.2.2, … 1.2.*n* (under stakeholder requirement 1.2), and so on. Such a traceability model allows for simple and obvious linkages back through the hierarchy of requirements.

[51] IEEE 29148-2018 clause 6.3.1

organizational, technical and physical environment. For an IoT solution, this is the environment in which the devices will operate, the organization implementing them, and the enabling systems (also called supporting systems or supporting services) that exist. Scenarios (use cases, user stories, etc.) may be developed to analyze the operation of the system in its intended environment.

c. *Constraints on the system are identified.* These may be imposed in the business requirements, derived from enabling systems with which the IoT solution must interface, and newly discovered constraints based on stakeholder needs. They may include budgetary constraints, regulatory constraints, and technical constraints imposed by existing systems.

d. *Stakeholder needs are defined[52] and translated into stakeholder requirements.* With the constraints defined and needs discovered, stakeholder requirements can be created. These should include requirements that are functional- and quality-related.

e. *Stakeholder needs are prioritized and transformed into clearly defined stakeholder requirements.* Stakeholder requirements analysis is performed to ensure the statements are constructed according to requirements engineering best practices (covered later in this chapter). It is also performed to ensure they are complete as a set (comprehensive) and prioritized. Stakeholder requirements should be necessary, implementation free[53], unambiguous, consistent, complete, singular, feasible, traceable, verifiable, affordable, and bounded.

f. *Traceability of stakeholder requirements to stakeholders and their needs is established and linked to business or mission requirements.*

---

[52] In some cases, two stakeholders or stakeholder groups will have opposing interests. Such opposition should be resolved so that stable requirements can be developed. They may be resolved through clear explanation of technical abilities within IoT solutions or by acquiring managerial input as to which opposing need is more important to the organization. Additionally, some stakeholders may oppose the IoT solution itself and these oppositions should be addressed through risk management processes. The stakeholder *need* of not implementing the solution will not be satisfied, but it may be addressed to accommodate their concerns with real solutions, even if they still oppose the IoT solution.

[53] Stakeholder requirements should not be written, just as business requirements, such that they limit the solution to a specific vendor, protocol, or technology.

3. **System [System/Software] Requirements Definition (expanded in 29148 and outlined in ISO/IEC/IEEE 15288:2015):** The purpose of this process is to transform the stakeholder, user-oriented view of desired capabilities into a technical view of a solution that meets the operational needs of the user. The system requirements define, from the supplier's perspective, the characteristics, and functional and performance requirements the system must possess to satisfy stakeholder requirements. These requirements should not imply a specific implementation (vendor, protocol, etc.), unless constrained to do so by a higher-level requirement or constraint. Outcomes include:
    a. *The system description, including functions and boundaries, is defined.*
    b. *System requirements (functional, non-functional, interface, etc.) and design constraints are defined.*
    c. *System requirements are analyzed to ensure proper construction and traceability to stakeholder and business or mission requirements and constraints.*

Figure 3.2a illustrates the scope of requirements and requirement processes and inputs.

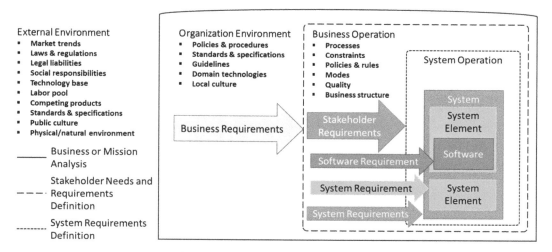

Figure 3.2a: Requirements Scope

The specifications that come from the requirements processes depend on the scope of the requirements. The lowest levels, system element or software requirements, are constrained by the stakeholder needs in business operations and the organizational environment as well as external influences.

29148-2018 defines Requirement Engineering in clause 5.2 as, an interdisciplinary function that mediates between the domains of the acquirer and supplier or developer to establish and maintain the requirements to be met by the system, software, or service of interest. Requirements engineering is concerned with discovering, eliciting, developing, analyzing, verifying (including verification methods and strategy), validating, communicating, documenting and managing requirements. The primary result of requirements engineering is sets of requirements, each set:

- being with reference to a defined system, software or service;
- enabling an agreed understanding between stakeholders (e.g., acquirers, users, customers, operators, suppliers);
- having been validated against real-world needs;
- able to be implemented; and
- providing a reference for verifying designs and solutions.

The above description of requirements engineering is our overall framework for creation of requirements. Figure 3.3 illustrates the interdisciplinary nature of the process and that which each party brings to the table.

The acquirer[54] is the stakeholder that acquires or procures a product or service from a supplier. It is the individual or group within the organization with the desire and authority to request and approve the development of a solution, in this case, a wireless IoT solution. The supplier is the organization, group, or individual that enters into an agreement with the acquirer to supply the product or service.

Both the acquirer and other intra-organizational and inter-organizational stakeholders must work together with the supplier to implement effective requirements engineering. The acquirer and other stakeholders bring vertical expertise to the process. By vertical expertise, we are referencing the business sector or group of similar organizations with

---

[54] Acquirers and suppliers may be internal or external. For example, an external acquirer may be an organization contracted by the target organization to manage the IoT solution project that in turn contracts with another organization as the supplier. The supplier may be internal when the organization has sufficient staff to perform the work or external when the internal staff is either technically incapable or unavailable due to time and workload constraints.

similar customers or group members for whom the acquiring organization operates, such as government, manufacturing, oil & gas, retail, hospitality, healthcare, entertainment, etc. The supplier brings technical expertise to the process, which, in the case of wireless IoT solutions, means an understanding of the IoT solution architectures, protocols, applications, and data processing.

While the acquirer domain provides the stakeholders, the supplier domain provides the technical professionals. The stakeholders have knowledge of the existing environment, including constraints and needs as well as future goals and objectives. The technical professionals have knowledge of IoT solutions, and the tools, planning, soft skills, and systems required for their implementation[55].

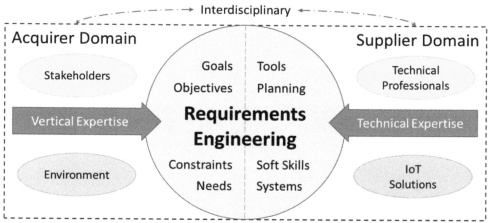

Figure 3.3: Interdisciplinary Requirements Engineering for IoT Solutions

3.2: Defining Stakeholders

Before the requirements can be engineered and documented the key stakeholders from each functional domain need to be established. The minimum stakeholders in the IEEE 29148 standard are referred to as users and acquirers. The complexity of the project will

---

[55] Proper requirements engineering requires valuable input from several areas and can be time-consuming, but a commitment to it results in far superior IoT solutions.

drive the number of stakeholders and how many different perspectives there are. It is common to uncover additional stakeholders during the process of creating requirements, this is a normal part of the iteration of requirements creation.

The 29148-2018 description of stakeholders is as follows[56]:

> Stakeholders vary across projects when considered in the context of requirements engineering. A minimum set of stakeholders consists of users and acquirers (who may not be the same). Complex projects can impact many users and many acquirers, each with different concerns. It may be necessary to include two other groups as part of the minimum set of stakeholders. First, the organization or organizations developing, maintaining or operating the system or software has a legitimate interest in benefiting from the system. Second, regulatory authorities can have statutory, industry, or other external requirements demanding careful analysis.

In small projects the stakeholder(s) are easy to identify. It is usually whoever is paying for the system and the immediate users. As the projects grow in scale and complexity the number of stakeholders tends to grow. If a solution already exists, the existing solution provider becomes a stakeholder. If the existing solution was developed by an outside organization, this kind of stakeholder would be an external stakeholder. In any case, defining the goals up front ensures alignment in reaching those goals and determining if they were obtained.

Sometimes it is helpful to have real world examples, for this, we will use a bill in California at various points in this chapter. The bill is known as California AB-1761 which covers employee safety for hotel workers. For more information on this bill, visit the California Legislature's website[57].

To summarize this legislation, the state of California proposed requiring hotels to provide their employees with a panic button to get help if they are in fear for their lives or safety. At the time of this writing, this bill has not become a state law; however, several municipalities did implement these rules.

---

[56] IEEE 29148-2018 clause 5.2.2

[57] leginfo.legislature.ca.gov/faces/billCompareClient.xhtml?bill_id=201720180AB1761

Listed below are each of the functional areas for the IoT system, along with the stakeholder listed after the area in the parentheses, for a solution that meets the AB-1761 requirements for a panic button.

- Organization that will benefit from and approve acquiring the IoT system (Hotel Owner)
- Primary users of the technology (Hotel Service Staff)
- Organization maintaining the IoT system (Hotel IT Staff)
- Organization maintaining the IoT network (Hotel IT Staff)
- Organization responsible for security of the IoT system and network (Hotel IT Staff and third-party provider)
- Organization developing the IoT system (third-party hardware/software provider)
- Regulatory Domain (FCC)
- Legal Authority who could have laws impacting the design (Local Government, State Government, Federal Government)

This is just a first pass at stakeholders (internal and external) and during the full requirements gathering process others may be identified.

For less complex projects, some of these may be combined. Other projects that are more complex could have additional stakeholders and groups. For example, more than one organization may benefit from the system resulting in multiple business owners. But this should give you an idea of who needs to have input into the system.

3.3: Turning Needs into Requirements

Now that we have the stakeholders identified. We need to turn their needs into requirements. 29148-2018 clause 5.2.3 introduces this concept of "Transformation of needs into requirements" as follows:

> Defining requirements begins with stakeholder needs (or goals, or objectives) that are refined and evolve before arriving as valid stakeholder requirements. Initial stakeholder concerns do not serve as stakeholder requirements, since they often lack definition, analysis and possibly consistency and feasibility. Using the

Concept of Operations[58] to aid the understanding of the stakeholder concerns at the organizational level and the System Operational Concept[59] from the system perspective, requirements engineering leads stakeholders from those initial, often latent, needs to a set of objectively adequate, structured and more formal statements of stakeholder requirements and goals.

In most cases, each stakeholder will have a goal, need, or objective from the IoT solution itself. These are similar to use cases in the Scrum/Agile project methodology. We will cover properly forming these requirements later in this section; however, it is important to remember that this is an iterative process. In an iterative process additional information may create a need to revisit a previous step. Recursion is also likely to occur. Iteration occurs when the same process is repeated at the same level of a system. Recursion occurs when the same process is revisited in successive or previous levels of a system. If we continue to use our hotel's panic buttons as an example, the purpose or primary goal of the system is for the primary users to have a solution to report a safety issue.

As such, the previous list of stakeholders is missing the secondary user of Hotel Security for the panic button system. The hotel security has a need-to-know the room or the location in which the individual who has pressed the panic button is located. This creates a requirement for location identification for the system or some method of tracking which button was pressed and where that button is located.

Requirements Construct

The language used in the creation of requirements is key to ensure alignment. Properly formed requirements need to be clear and unambiguous so that stakeholders understand

---

[58] A *Concept of Operations* is the user definition of how the overall organization will be operated to satisfy its mission. It is a verbal and graphic statement, in broad outline, of an organization's (enterprise's) assumptions or intent in regard to an operation or series of operations of new, modified, or existing organizational (enterprise) systems. *Definition from ANSI/AIAA G-043B-2018.*
[59] The *System Operational Concept,* or simply *Operational Concept,* is the user definition of how a specific system will be utilized within the organization. It may include a flow down of the concept of operations activities to be performed using the specific system and/or a verbal and graphic statement of an organization's (enterprise's) assumptions or intent in regard to an operation or series of operations of a specific system or a related set of specific new, existing, or modified systems. *Defined in ANSI/AIAA G-043B-2018.*

each requirement. These requirements must also be able to be validated. How can there be an agreement between stakeholders if the language itself is vague, ambiguous, or unclear? How can the requirements be implemented or validated if people have different or even conflicting understanding of the requirements themselves? The short answer to these questions: you cannot be successful without a proper construct of the requirements because you do not know how to measure success.

This makes ensuring requirements are properly formed with a common understanding for each requirement of the utmost importance. Properly formed requirements will eliminate ambiguity, provide clarity on what is or is not a requirement for the design and implementation of the IoT network, and identify any constraints. With so much depending on creating well-formed requirements constructs, it is best to go directly to the standard. The 29148-2018 team worked to define a methodology and the proper language to use evolved over 49 years based on underlying standards.

The 29148-2018 details the requirements construct with guidelines for structuring statements so that they are clear and unambiguous[60].

> Well-formed stakeholder requirements, system requirements and system element requirements shall be developed. This practice contributes to requirements validation with the stakeholders and helps ensure that the requirements accurately capture stakeholder needs.
>
> A well-formed specified requirement contains one or more of the following:
>
> - it shall be met or possessed by a system to solve a problem, achieve an objective or address a stakeholder concern;
> - it is qualified by measurable conditions;
> - it is bounded by constraints;
> - it defines the performance of the system when used by a specific stakeholder or the corresponding capability of the system, but not a capability of the user, operator, or other stakeholder; and
> - it can be verified (e.g., the realization of the requirement in the system can be demonstrated).

---

[60] IEEE 29148-2018 clause 5.2.4

This description provides a means for distinguishing between requirements and the attributes of those requirements (conditions, assumptions and constraints).

It is important to agree in advance on the specific keywords and terms that signal the presence of a requirement. A common approach is to stipulate the following.

- Requirements are mandatory binding provisions and use 'shall'.
- Non-requirements, such as descriptive text, use verbs such as 'are', 'is', and 'was'. It is best to avoid using the term 'must', due to potential misinterpretation as a requirement.
- Statements of fact, futurity, or a declaration of purpose are non-mandatory, non-binding provisions and use 'will'. 'Will' can also be used to establish context or limitations of use.
- Preferences or goals are desired, non-mandatory, non-binding provisions and use 'should'. They are not requirements.
- Suggestions or allowances are non-mandatory, non-binding provisions and use 'may'.
- Use positive statements and avoid negative requirements such as 'shall not'.
- Use active voice: avoid using passive voice, such as 'it is required that'.
- Avoid using terms such as 'shall be able to'.

All terms specific to requirements engineering should be formally defined and applied consistently throughout all requirements of the system.

The template for a requirement statement, at least a functional requirement statement, is defined in the 29148-2018 standard as:

[Condition][Subject][Action][Object][Constraint of Action]

For example:

When an IoT device sends a message to the network **[Condition]**, a gateway **[Subject]** shall be available **[Action]** to receive the message **[Object]** within all areas of the manufacturing plant [**Constraint of Action**].

Conditions are measurable qualitative or quantitative attributes that are stipulated for a requirement. They further qualify a requirement that is needed and provide attributes that permit a requirement to be formulated and stated in a manner that can be validated and verified. Conditions may limit the options open to a designer. It is important to transform the stakeholder needs into stakeholder requirements without imposing unnecessary bounds on the solution space.

Constraints restrict the design solution or implementation of the systems engineering process. Constraints may apply across all requirements, may be specified in a relationship to a specific requirement or set of requirements, or may be identified as stand-alone requirements (i.e., not bounding any specific requirement).

Examples of constraints giving rise to requirements include:

- interfaces to already existing systems (e.g., format, protocol or content) where the interface cannot be changed;
- physical size limitations (e.g., a controller shall fit within a limited space in an airplane wing); — laws of a particular country;
- available duration or budget;
- pre-existing technology platform;
- maintenance constraints; or
- user or operator capabilities and limitations.

Requirements may be ranked or weighted to indicate priority, timing or relative importance. Requirements in scenario form depict the system's action from a user's perspective.

Using the above methodology to construct requirements is the best practice[61]. Implementing this framework is time intensive; however, it ensures that wireless IoT solution will meet the requirements because it ensures that the requirements are well defined.

---

[61] IEEE 29148-2018 Clause 5.2.4

29148-2018 goes on to define characteristics of individual requirements[62] which serve as a great checklist or passthrough filter for determining if a requirement is good or bad as constructed. Here is the summary checklist recommended for each requirement:

- Is the requirement necessary?
- Is the requirement appropriate?
- Is the requirement unambiguous?
- Is the requirement complete?
- Is the requirement singular?
- Is the requirement feasible?
- Is the requirement verifiable?
- Is the requirement correct?
- Is the requirement conforming?

After writing each requirement, evaluate each against these questions. This is an important step in the process to ensure a proper requirement construct. Failure to meet any of these criteria may lead to confusion, frustration, and possibly failure to meet that requirement. If an individual requirement fails to meet one of these criteria, the requirements engineer can modify the requirement itself and get agreement from the stakeholders of the project that the intent is met with these changes.

These questions are based on the following term definitions[63]:

- **Necessary:** The requirement defines an essential capability, characteristic, constraint and/or quality factor. If it is not included in the set of requirements, a deficiency in capability or characteristic will exist, which cannot be fulfilled by implementing other requirements. The requirement is currently applicable and has not been made obsolete by the passage of time. Requirements with planned expiration dates or applicability dates are clearly identified.
- **Appropriate:** The specific intent and amount of detail of the requirement is appropriate to the level of the entity to which it refers (level of abstraction appropriate to the level of entity). This includes avoiding unnecessary constraints on the architecture or design while allowing implementation independence to the extent possible.

---

[62] IEEE 29148-2018 clause 5.2.5
[63] IEEE 29148-2018 clause 5.2.5

- **Unambiguous:** The requirement is stated in such a way so that it can be interpreted in only one way. The requirement is stated simply and is easy to understand.
- **Complete:** The requirement sufficiently describes the necessary capability, characteristic, constraint, or quality factor to meet the entity need without needing other information to understand the requirement.
- **Singular:** The requirement states a single capability, characteristic, constraint, or quality factor. Although a single requirement consists of a single function, quality, or constraint, it can have multiple conditions or constraints under which the requirement is to be met.
- **Feasible:** The requirement can be realized within system constraints (e.g., cost, schedule, technical) with acceptable risk.
- **Verifiable:** The requirement is structured and worded such that its realization can be proven (verified) to the customer's satisfaction at the system level where the requirement exists. Verifiability is enhanced when the requirement is measurable.
- **Correct:** The requirement is an accurate representation of the entity need from which it was transformed.
- **Conforming:** The individual items conform to an approved standard template and style for writing requirements, when applicable.

Beyond the Exam: Historic Views of Requirement Statements

On the topic of requirement characteristics, consider that the importance of these characteristics goes beyond the individual requirements. Ultimately, you are creating a system requirements specification, or SRS, and the SRS should comply with these characteristics. From the British Telecommunications Engineering journal, Volume 6, Part 4 supplement from 1988[64], the following statements hold true today:

An SRS is unambiguous only if every requirement described in it is unambiguous. A requirement is unambiguous if it has only one possible semantic interpretation.

---

[64] This supplement is available for free access on archive.org at archive.org/details/bte-supplement-198801. The supplement is 16 pages in length and provides additional useful information both similar and dissimilar to that in the IEEE 29148-2018 standard.

An SRS is complete if it states the full set of the customer's requirements, and each requirement has been stated in full.

An SRS is correct if every requirement stated in it has been verified by the customer.

An SRS is consistent if there is no conflict between any given set of requirements described in it.

An SRS is verifiable if every requirement stated in it is verifiable. A requirement is verifiable only if there exists some process by which the developed system can be checked to ensure that the requirement has been fulfilled.

An SRS is modifiable if its structure and style are such that any necessary changes to the requirements can be made easily, completely, and consistently.

The requirements in an SRS are traceable if the origin of each of its requirements is clear and if the SRS facilitates the referencing of each requirement.

While these listed characteristics do not match those in the IEEE standard exactly, they are similar and provide additional insight. Additionally, the supplement makes the following statement, which reveals the reality of requirements specification development, even more than thirty years later, "It would be remarkable if an SRS achieved all these attributes perfectly, since customers' requirements are arrived at through evolution rather than spontaneity. However, the use of guidelines forces the customer to consider, in the first instance, many points which may otherwise have been deferred until later, and it thus channels and accelerates the process of SRS production." This comment shows both the reality of requirements definition, even with a standard process, and the benefit of using a standard process in spite of the fact that some requirements will evolve throughout design and deployment.

Finally, the referenced supplement provides an excellent template for a requirements specification document. Each section is clearly explained with checklists for each section. Do not be fooled by the age of this supplement and its free availability on archive.org. The document is still an excellent source of information for requirements development in modern IoT projects. We either learn from the past or we pass on learning.

-Tom

Requirements Set Construct

Creating each individual requirement is important; however, the requirements need to come together as a requirements set. The 29148-2018 document covers in detail how to validate your requirements as a requirement set in clause 5.2.6. The questions that need to be answered against a requirements set are simply:

- Are the requirements complete?
- Are the requirements consistent?
- Are the requirements feasible?
- Are the requirements comprehensible?
- Are the requirements able to be validated?

All these questions are equally important and together help ensure the requirements set is obtainable. Failure to meet any of these criteria can lead to confusion, frustration, and possibly failure of the IoT solution. Imagine if you have two requirements that are not consistent and conflict with each other? Or an incomplete set of requirements? Neither of these situations is good for the overall project.

If the requirements set fails to meet one of these criteria, requirements may need to be added, deleted, or modified. In this step it is common to find missing requirements and/or additional stakeholders. The goal here is to lock down the requirements set to ensure there is no creep of requirements, as creep jeopardizes the success of the overall project.

These questions are based on the following term definitions[65]:

- **Complete:** The set of requirements stands alone such that it sufficiently describes the necessary capabilities, characteristics, constraints or quality factors to meet entity needs without needing further information. In addition, the set does not contain any To Be Defined (TBD), To Be Specified (TBS), or To Be Resolved (TBR) clauses. Resolution of the TBx designations may be iterative and there is an acceptable timeframe for TBx items, determined by risks and dependencies. To improve completeness, the following practices can be adopted:

---

[65] IEEE 29148-2018 clause 5.2.6

- include all requirements types relevant to the system under consideration;
- account for requirements in all stages of the life cycle; and
- involve all stakeholders in the requirements elicitation, capture, and analysis activity[66].

- **Consistent:** The set of requirements contains individual requirements that are unique, do not conflict with or overlap with other requirements in the set, and the units and measurement systems are homogeneous. The terminology used within the set of requirements is consistent, i.e., the same term is used throughout the set to mean the same thing.
- **Feasible:** The complete set of requirements can be realized within entity constraints (e.g., cost, schedule, technical) with acceptable risk. Feasible includes the concept of 'affordable'.
- **Comprehensible:** The set of requirements is written such that it is clear as to what is expected by the entity and its relation to the system of which it is a part.
- **Able to be validated:** It is practicable that satisfaction of the requirement set will lead to the achievement of the entity needs within constraints (e.g., cost, schedule, technical, legal, and regulatory compliance).

Careful validation of the requirements set against these characteristics is critical to avoiding requirements changes and growth ('requirements creep') during the life cycle that will impact the cost, schedule, or quality of the system.

Language to Avoid

As we mentioned before, business requirements should focus on *what* needs to be accomplished, not *how* these tasks will be completed. Technical requirements will be created to satisfy the business requirements as part of an iterative requirements creation process. Keeping these two types of requirements separate can be difficult to do; however, focusing on the business requirements first can allow for multiple different technical solutions to fill the needs. Each requirement will go through iterations until the business and technical requirements are finalized.

---

[66] It is common to need to include TBx designations during the evolution of the requirements definition, as the process is informed by system analysis results and trade-off decisions. However, the set of requirements cannot be considered complete until all the TBx designated requirements have been resolved.

Just as the construct of the requirements is important, so is the language used to form those requirements. Statements must be clear and unambiguous to each of the stakeholders and the design and deployment teams. If a word can have more than one meaning, make sure to clarify and document the intended use with a definition statement in the requirements document. In this section we will cover language to avoid.

To create clear and unambiguous requirements, it is best to avoid generic terminology. Ambiguous terms can be adverbs, adjectives, *if* logical statements, or pronouns. A statement like *the IoT system must function with minimal downtime* is ambiguous and not measurable. Whereas a statement like *the IoT system shall function as defined by the requirements in this document with 99.9 percent uptime during operations hours* is a much better requirement assuming the readers know or have a source to discover the meaning of *operations hours*. This requirement is measurable and obtainable as a result.

Subjective language is another area to avoid for similar reasons. What does *cost effective* or *user-friendly* mean? A better way to communicate is to allow for a maximum budget for implementation and operations instead of stating that it should be cost effective. Instead of *user-friendly*, define requirements for the usability and quality of the system itself. How long should a task take to be accomplished? What kind of interface should be provided? What HCI (human-computer interaction) guidelines and methodologies will be followed? What usability testing methods will be employed?

Comparative phases are similar in nature, phrases like *better than* or *higher quality* are not able to be measured and thus make poor requirement statements. Yes, we all want something to be high quality and better than the legacy system, but the meaning needs to be defined among the stakeholders in a measurable way, else, it cannot be validated. Instead of *higher quality* we may be trying to increase the mean time between failures (MTBF) from 2 years to 3 years. Instead of *better than* a performance goal of performing a task in 100 milliseconds instead of 200 milliseconds may be appropriate.

We must ensure all requirements are verifiable; that the requirements can be validated in an implemented wireless IoT solution. When creating the requirement, think about how it will be verified when the project is complete. If there isn't a way to validate the requirement, consider removing the item as a requirement and change it to a desired feature or capability or discover a way to define it which is verifiable.

Other common problems are confusing words. *Must* is a common confusing word that gets inserted into requirements; however, *shall* or *should* are better word choices offering more flexibility and accuracy depending on the goals of the requirement. If you use the word must, define its meaning in your document. For that matter, define all constraining or recommending words, such as must, shall, should, may, can, and optionally as a qualifier. You can also borrow from standards documents. For example, the IEEE 802.11-2020 standard states:

> In this document, the word *shall* is used to indicate a mandatory requirement. The word *should* is used to indicate a recommendation. The word *may* is used to indicate a permissible action. The word *can* is used for statements of possibility and capability[67].

The Achilles heel of requirements is utilizing loophole statements. Examples of utilizing a loophole are terms like *as appropriate* or *if possible*. Each of these leaves an escape clause for the requirement when the project is finished. Imagine being finished with a project and one of your core items isn't working and the system developer says, "Oh, that wasn't possible". Leaving requirements with these statements keeps too much open to interpretation and creates a challenge for the engineer implementing the requirements.

Terms that imply totality should also be avoided, as they can create validation problems. How do you test a condition like *always* or *never*? The testing and QA cycles would endure to infinity, else, you could not be sure something *never* occurred or *always* occurred. The goal is to have a reliable system, but to implement something that *never* has an outage is impossible for even the best design professionals.

Any references to other requirements or outside sources should be complete. If you reference another requirements document or requirement statement, make sure you specify the version number and/or date of that document or requirement statement. This is an iterative process, so the upstream requirement may change and leave a downstream requirement stranded.

A quick checklist to utilize along with examples of language to avoid within the requirements is available in Table 3.1.

---

[67] IEEE 802.11-2020 clause 1.4 Word Usage

Things to avoid	Examples
Non-Requirements	Are, Is, Was
Confusing Terms	Must
Superlatives	Best, Most
Subjective Language	User Friendly, cost effective
Vague Pronouns	It, This, That
Ambiguous Terms	Almost always, Significant
Non-Verifiable Terms	Provide support, as a minimal
Comparative Phrases	Better than, Higher Quality
Loopholes	If possible, as appropriate
Terms that Imply Totality	All, always, never, every
Incomplete References	Indicating a document statement without giving enough detail to locate the document

Table 3.1: Language to Avoid

Clause 5.2.7 provides an overview of the terms and phrases listed in Table 3.1 and explained above. The clause is summarized with the following important statement about assumptions:

> All assumptions made regarding a requirement shall be documented and validated in one of the requirement's attributes (e.g., rationale) associated with a requirement or in an accompanying document. Include definitions as declarative statements, not requirements.

For each requirement statement, a set of attributes, as referenced in the preceding quote, can be assigned to that statement to allow for requirements tracking throughout the solution design, deployment, and validation life cycle. These attributes include[68]:

- **Identification:** Each requirement should be uniquely identified (i.e., number, name tag, mnemonic). Identification can reflect linkages and relationships, if needed, or they can be separate from identification. Unique identifiers aid in requirements tracing. Once assigned, the identification is unique - it is never changed (even if the identified requirement changes) nor is it reused (even if the identified requirement is deleted).

---

[68] IEEE 29148-2018 clause 5.2.8.2

- **Version Number:** This is to make sure that the correct version of the requirement is being implemented as well as to provide an indication of the volatility of the requirement. A requirement that has been changed many times could indicate a problem or risk to the project.
- **Owner:** The person or element of the organization (for example, a department) that maintains the requirement, who has the right to say something about this requirement, approves changes to the requirement, and reports the status of the requirement.
- **Stakeholder Priority:** The priority of each requirement should be identified. This may be established through a consensus process among potential stakeholders. As appropriate, a scale such as 1-5 or a simple scheme such as High, Medium or Low, could be used for identifying the priority of each requirement. The priority is not intended to imply that some requirements are not necessary, but it may indicate what requirements are candidates for the trade space when decisions regarding alternatives are necessary. Prioritization needs to consider the stakeholders who need the requirements. This facilitates trading off requirements and balancing the impact of changes among stakeholders.
- **Risk:** A risk value assigned to each requirement based on risk factors. Requirements that are at risk include requirements that fail to have the set of characteristics that well-formed requirements should have. Failing to have these characteristics can result in the requirement not being implemented (fails system verification) and entity needs not being realized (fails system validation). Risk can also address feasibility/attainability in terms of technology, schedule, cost, politics, etc. If the technology needed to meet the requirement is new with a low maturity the risk is higher than if using a mature technology used in other similar projects. Risk may also be inherited from a parent requirement.
- **Rationale:** The rationale for establishing each requirement should be captured. The rationale provides the reason that the requirement is needed and points to any supporting analysis, trade study, modelling, simulation or other substantive objective evidence.
- **Difficulty:** The assumed difficulty for each requirement should be noted (e.g., Easy/Nominal/Difficult). This provides additional context in terms of requirements breadth and affordability. It also helps with cost modelling.

- **Type:** Requirements vary in intent and in the kinds of properties they represent. Use of a type attribute aids in identifying relevant requirements and categorizing requirements into groups for analysis and allocation.
    - *Functional/Performance:* Functional requirements describe the system or system element functions or tasks to be performed by the system. Performance is an attribute of function. A performance requirement alone is an incomplete requirement. Performance is normally expressed quantitatively. There can be more than one performance requirement associated with a single function, functional requirement, or task.
    - *Interface:* Interface requirements are the definition of how the system is required to interact with external systems (external interface), or how system elements within the system, including human elements, interact with each other (internal interface). External interface requirements state characteristics required of the system, software or service at a point or region of connection of the system, software, or service to the world outside of the item. They include, as applicable, characteristics such as location, geometry and what the interface is to be able to pass in each direction.
    - *Process Requirements:* These are stakeholder, usually acquirer or user, requirements imposed through the contract or statement of work. Process requirements include compliance with national, state, or local laws, including environmental laws, administrative requirements, acquirer/supplier relationship requirements and specific work directives. Process requirements may also be imposed on a program by corporate policy or practice. System or system element implementation process requirements, such as mandating a particular design method, are usually captured in project agreement documentation such as contracts, statements of work and quality plans.
    - *Quality (Non-Functional Requirements):* Include a number of the 'ilities' in requirements to include, for example, transportability, survivability, flexibility, portability, reusability, reliability, maintainability and security. The kinds of quality requirements (e.g., "ilities") should be identified prior to initiating the requirements activities. This should be tailored to the system(s) being developed. As appropriate, measures for the quality requirements should be included as well.

- *Usability/Quality-in-Use Requirements:* For user performance and satisfaction, they provide the basis for the design and evaluation of systems to meet the user needs. Usability/Quality-in-Use requirements are developed in conjunction with, and form a part of, the overall requirements specification of a system and are discussed later in this chapter in greater detail.
- *Human Factors Requirements*[69]: State required characteristics for the outcomes of interaction with human users (and other stakeholders affected by use) in terms of safety, performance, effectiveness, efficiency, reliability, maintainability, health, well-being, and satisfaction. These include characteristics such as measures of usability, including effectiveness, efficiency, and satisfaction; human reliability; freedom from adverse health effects.

3.4: Use Cases and Solution Justification

Now that we have the language, construct, and criteria around requirements defined we can move onto forming a requirements set. We need to step back and explore use cases and solution justification first. For the purposes of this book, we will continue to use California AB-1761 as our subject to create requirements. We can match each requirements section to the law and build out a model for a requirements document. This will be a sample and is intended to show the process and not to be utilized in lieu of actual gathering requirements for the project or from the customer.

The first step to creating requirements is understanding the use cases and what the business needs are. With business requirements these are the business drivers that generate the need for an IoT solution, whether wired or wireless. It could be the building owner, with a goal like saving on electricity. This goal could drive the need for an energy management system to track energy usage and adjust the environment to reduce usage, thus costs. Or possibly a farmer who wants to conserve water. This need might demand

---

[69] Within the 29148-2018 definitions, the primary difference between the Quality Requirements and the Human Factor Requirements types is that quality is focused on the factors of the system or solution and human factors is focused on the people using or in the environment of the system or solution.

an IoT network of sensors to report on moisture levels of the soil at different locations that will drive the watering of those areas while avoiding overwatering others.

In the case of our example, having the panic button is potentially a legal requirement. Failure to provide this system will result in a legal liability resulting in daily fines. In addition to the legal ramifications, the business may have issues with staff if they feel they are not safe. So, providing the system makes great business sense for them.

Here is the actual legal requirement:

> (1) Provide employees working alone in a guestroom with a panic button, free of charge. The employee may use the panic button, and cease work, if the employee reasonably believes there is an ongoing crime, harassment, or other emergency happening in the employee's presence. The hotel employer shall develop an appropriate protocol, including any necessary training, for how staff, security, and management shall respond when a panic button is activated. The protocol shall be calculated to ensure an immediate on-scene response to the greatest extent possible.

The use case from the bill is to provide specific employees, those working alone in a guestroom, with a panic button to use if they believe there is an ongoing crime or harassment. The justification for providing this panic button is employee safety and the fact that this bill is required in some municipalities.

3.5: Coverage Area Requirements

Earlier in this chapter, we addressed requirements in general, how to create requirements that have meaning, can be measured, etc. This structure can be used for anything from designing and developing a product to deploying the network. In this section we will discuss common business requirements for designing systems that will be used in the real world.

For purposes of writing requirements, the coverage area(s) for the wireless IoT solution needs to be defined. Often, we define the coverage area(s) for stationary systems and specific areas to exclude from coverage. The designer may use color coding for areas of high user density, high bandwidth demands, or just a coverage and define them as requirements areas. This can be helpful when designing for an IoT network, to visually show the goals of the network.

Different IoT systems and business models will have different coverage requirements. In cases of a wholly owned and operated network, the system may be limited to the space used by the business that is purchasing the network. City-owned or public systems may want to cover an entire metro area. Most deployments covering an entire metro area utilize low bandwidth technologies, such as LoRaWAN and Sigfox, as it may be cost prohibitive to deploy network technologies with limited coverage ranges, such as 6LoWPAN, Zigbee, and WirelessHART.

In our hotel example, the IoT system needs to provide connectivity within the guest rooms of the hotel. A secondary requirement is for there to be a response to the panic button because determining the location of the user is also important from a business perspective. Therefore, this seemingly simple need, having a panic button, quickly becomes a complex wireless IoT solution involving wireless network protocols, location services, and applications that can receive the data and trigger a response (through alerts to appropriate individuals and possibly law enforcement). Having broad coverage with few gateways or access points for the IoT system may be counterproductive for the system to determine the location of the user. Additionally, one technology may be used for locationing, like BLE, while other technologies are used for transmitting information among systems like Wi-Fi, Zigbee, MQTT, and more. Of course, if you are only implementing the panic button solution, it could be accomplished most easily with just one wireless protocol and then higher layer application architectures that get the information to the right place quickly.

It is just as important to document any areas that are to be specifically excluded. These are areas where coverage should not exist. And anything outside of the specific coverage or exclusion areas is a grey area of "don't care" if coverage bleeds there. To be clear, it is very challenging to truly prevent coverage in an area near the gateways or mesh routers in a wireless IoT network. The "coverage" is there if the end device can see and connect to the network, and this is a factor of the end device's receive sensitivity and antenna gain as well as the transmit power and antenna gain of the gateways or mesh routers. However, you can limit the range where a typical receiver can indeed participate in the network or eavesdrop on the network.

With IoT systems that require mobility of the system, like Personal Area Network (aka PAN), the coverage area may move with the users. PAN networks tend to be low powered and operated from a device that is carried with the person. A good example of these types of systems is to think of a cell phone providing a Bluetooth connection to a FitBit™ and utilizing either 5G or Wi-Fi to backhaul the data from this device.

During business and stakeholder requirements definition, you will identify coverage areas and requirement areas. During technical requirements specification, you will identify the technical requirements of the wireless IoT solution that will accomplish this goal.

3.6: System Users

It is also important for the business to define who will be using the wireless IoT solution and how will they use it. The term IoT is wide ranging and general. The term is so broad one could not be more specific than *things*. The number of different verticals of *things* is massive, but the ways each of these verticals use IoT systems is just as diverse. In addition to this, who is using the system and how they are using it varies by industry. The way we think about how users will interact with the energy management system is very different from how they would interact with an IoT system in agriculture. Both of those systems are also vastly different from medical monitoring systems, which may actually be providing life-saving information to the medical team.

With many IoT systems the primary use of the system is to collect data which is then used to inform business decisions. In the case of agriculture, the farm owner may receive data that the soil needs water to provide the optimal growing environment for the crops that are currently planted. Identifying the user in this type of system is simple because it is a small subset of users.

In the hotel panic button example, there are two primary users of the system. First the service staff who initiate the alert and secondly the security team which responds to the alert. Other secondary users of the system may include the hotel owner or management and law enforcement in the event there is a crime being committed and charges are filed.

Use cases indicate coverage requirements, or where the solution must be available for use, and suggest the users of the solution. Users indicate capability requirements or what the solution must be able to do for those users.

3.7: System Interfaces and Integration

Defining interfaces an IoT solution will have with other systems and within its own designed architecture cannot be an afterthought. Interfaces tend to fall into one of three primary categories.

- Interfaces to back-end systems (external or internal interfaces for the IoT solution)
- User Interfaces (internal interfaces for the IoT solution)
- Interfaces to other systems (external interfaces for the IoT solution)

It can be helpful to diagram the key components of the system before writing these requirements. Figure 3.4 shows a sample diagram for a vending machine IoT solution proposed by Intel for an IoT retail gateway reference design.

The IoT solution uses an IoT gateway to connect to cloud-based vending services and business management. Internal to the vending machine is payment processing, temperature management for refrigeration, vending product deliver mechanisms, and other platform sensors (such as inventory level detection). The human interface is provided through touch screen sensors and an advertising screen. The solution defines a vending API for cloud connectivity. An IoT solution as simple as a vending machine system can become quite complex when each of hundreds or thousands of vending machines is considered an individual or collection of IoT end devices that integrates with a cloud architecture. Interfaces include those among the components within the vending machine, those between the components and the gateway, and that between the gateway and the cloud

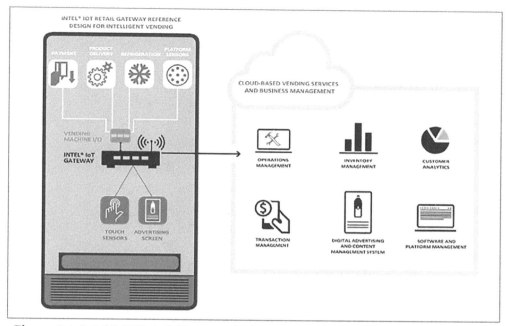

Figure 3.4: Intel IoT Retail Gateway Reference Design for Intelligent Vending[70]

The three interface categories previously listed are described briefly in the following sections.

Interfaces to Back-End Systems (Thing-Related Services)

It is common for IoT systems to utilize IoT sensors deployed across the network which will communicate back to a centralized server or to a centralized service that may be comprised of multiple servers and serverless functions[71]. This centralized component may be on-premises or hosted in a cloud environment run by the software provider. If you are designing the end-to-end IoT system you will have a lot of flexibility in these

---

[70] Intel® IoT Retail Gateway Reference Design for Intelligent Vending: Product Brief
[71] Serverless functions are quite popular in cloud technologies. The function is defined and stored within the cloud and the cloud service launches the function upon request (on any capable server), performs the proper actions, returns the necessary results, and removes the function from memory. Lambda functions in AWS, Azure functions in Microsoft Azure, and Serverless Compute in Google Cloud services are all examples of serverless functions.

interfaces. If you are purchasing an off-the-shelf or vendor-customized solution, these interfaces may be much more rigid in how they operate.

For business requirements, the location and needs for each server or service should be defined. Is the technical team building something on-site or off-site? Will there be an Application Programming Interface (API) between systems, or will the data be transferred to a location and processed on demand (a kind of queueing system)? What kind of information will be accessible between these systems?

The frequency at which data is available and how communications between the sensors and the back-end systems are also important to understand. When we begin defining the technical requirements these will be defined in more detail including any interfaces between sensors and backend systems that utilize the network. This should include the protocol being used, which sockets or ports on which the communications will take place, the hosts with which these communications will occur, etc. Of importance to the back-end systems is understanding if there is a requirement for Internet access as well. For example, cloud based IoT services require Internet access, but they are not the only ones. Some organizations implement Internet-based servers for distributed IoT solutions, and the servers are accessed by the IoT end devices or gateways. The servers are simply hosted by a hosting provider or accessible in the organizations DMZ with no cloud services involved[72] either on the Internet (public) or on-premises (private).

User Interfaces

Thought and planning needs to go into how end users will interact with the IoT systems being designed. Sometimes the IoT system will integrate into an electronic device the users already have, in other cases a dedicated device is required. At other times, the users have a personal or shared dashboard on a computer used to monitor and interact with the IoT solution. When gathering requirements for the user interface it is important to think about whether each set of users may have the same or different interfaces. The timeliness of the interaction needs to be outlined within these requirements as well.

Going back to the panic button example, the house keeping team's interface is a button, whereas the security team will get an immediate alert notifying them of the issue with

---

[72] Remember, it's only cloud if you don't know where it's at. Cloud services abstract the physical location from consideration whether they are private or public clouds.

the location of the individual, likely on a cell phone or another kind of paging device. In this case, failure to provide this alert in a timely fashion creates a serious problem for everyone involved. In fact, we would not even suggest that the requirement state that *the security team must receive an immediate alert*. Instead, state *that the security team must receive an alert within two seconds of the panic button press* or another specific time interval. When technical requirements are developed, this stakeholder quality requirement will assist in determining the wireless architecture used and the IoT application architecture as well.

User interfaces can take many forms. On the simple side of the scale a green or red light may be used to let the user know if the system is functioning. On the more complex side, the system may require a dedicated monitoring console that is custom made to operate the system. There is not a right or wrong user interface for IoT solutions; however, there is a right or wrong interface for a specific IoT solution. So, collecting an accurate and full set of requirements ensures a successful project.

When defining how the users will interact with the IoT solution, it is also important to define how often users will be interacting with the system. Fully autonomous systems may not require regular interaction, they may just alert when an exception occurs. Other systems may require regular monitoring or continuous monitoring depending on the use cases for the IoT solution.

Interfaces to Other Systems (Existing or In-Development)

In many cases IoT systems need to talk to other systems. It is extremely common for IoT systems to bring in automation use cases to trigger actions related to other systems, which may or may not be IoT-related themselves. These interfaces could be as simple as sending an email or logging the event. In other more complex cases, the system may have a requirement to record video any time a door is opened or closed and to analyze the video using AI/ML algorithms to determine if the individual(s) are known or unknown and if the behavior is normal or abnormal for the area. Or maybe you only want to record video if a smoke or vapor detector is triggered. The beauty if IoT solutions is also the complexity: many options are available.

Interfaces to other systems may be required for already existing systems or systems in-development. For both system types, if the interface to them cannot be changed and it already exists, the requirement to interface with them imposes a technical constraint. If the interface to them can be changed or is not yet defined, interfacing with them imposes a technical requirement.

Defining the interfaces that the business needs is important to set forth in the initial requirement engineering process. Additional business processes or needs can be added with iterations of the requirements as well. Diagraming the logical flow of these needs will be helpful as well.

Defining the details of how the interface to these other systems will work falls under the technical requirements. There are several technical details that need to be worked out when interfacing with other systems.

3.8: IoT System Data and Integration

In most cases, the value to the business resides in the data being collected by the IoT system. Data can be used to inform decisions by the business. Stadiums or arenas may implement an IoT system to give the people attending events wayfinding information to navigate the facility. This same system can be utilized to determine how many people are congregating in each area or that pass by an area. These stadium analytics can be utilized by the arena to optimize placement of merchandise, concessions, or advertising among other things.

This is one of the endless examples of the IoT system bringing value to all the stakeholders; however, the specific examples for the system being designed need to be created as part of this process. Here are some basic questions to think about when creating requirements around IoT system data:

- What data is valuable to the business?
- What data is valuable to the customer?
- How can this data be used to be more efficient or reduce expenses?
- Where is this data needed?
- When is this data needed?
- Who requires access?

In the case of commercial IoT solutions, the customer has access to the data and the solution provider may also have access to the data. In these cases, the solution provider can modify and improve the product based on how the end users are using it.

With the data being this important, when creating the IoT system requirements, it is important to define what data is essential versus desired versus irrelevant. This decision will enable the teams creating the technical requirements to ensure the system collects the correct data to bring value to the business owners.

The concept covered here is data and event collection and control. Given that IoT is about controlling machinery and systems or monitoring them or the environment, it's ultimately all about the data and events. If you're not getting the right data from sensors, the project fails. If you're not getting the right commands to actuators, the project fails. Therefore, defining system data and integration to allow for data and event collection and control is essential

An additional component of data integration that should be part of your design is the sharing of data among multiple IoT solutions or between IoT solutions and non-IoT systems. Some legacy systems may be able to use the data from the IoT solution but require that it be structured and formatted to the requirements of those systems. These scenarios will require traditional extract, transform, and load (ETL) procedures to perform data transformations so that the data is properly structured for the target legacy system. Once transformed, it can be loaded into the database system used by the legacy system.

The same is true in reverse. The IoT solution may make better decisions when data from the IoT solution is coupled with data from existing legacy systems. The process required to perform ETL is documented in detail in the CWIIP certification materials. For the wireless IoT solutions designer, it is essential only to know that the capability exists and to specify it as required, when necessary, within an IoT solution design.

3.9: Common Constraints

A constraint is defined as an externally imposed limitation on the system, its design, or implementation or on the process used to develop or modify a system. A constraint is a factor that is imposed on the solution by force or compulsion and may limit or modify the design[73]. Common constraints include aesthetic, architectural, budgetary, industrial, legal, regulatory, and timing. Each of these come with their own unique challenges and need to be addressed accordingly.

Aesthetic and Architectural Constraints

In some IoT systems there is no need for the sensors to meet any particular look or feel. In areas that are public facing it is very common to require sensors to either be out of sight or hidden in so that they do not create an unpleasant visual experience. Disney, for example, wants everything to blend in with its parks, so they have sensors inside ball heads camouflaged with the iconic mouse logo, as seen in Figure 3.5.

Figure 3.5: Aesthetic Design at Disney

With aesthetic constraints, there may be a specific size requirement or a visual requirement on the look-and-feel of the solution. Different organizations will have different requirements for these constraints; however, communicating what the expectations are early in the project can avoid a failure in meeting these expectations. Many want the IoT solution to do its work, but they don't want to see it doing its work in the physical world. Others will not be concerned about aesthetics at all because the IoT solution will be implemented in an environment already littered with machinery and legacy control devices.

---

[73] IEEE 29148-2018 clause 3.1.6

An example of an architectural constraint could be keeping a temperature sensor on a drone under 100 grams and no more than 2 x 2 x 2 centimeters so that the drone can maintain its ability to fly. You see, in the IoT space, things are connected to the IoT network through a wireless (or wired) module and have sensor or actuators involved that allow for monitoring and control of the thing. Each implementation must consider where the IoT device can be mounted on the thing, how it will interact with the thing, and how it will be powered by the thing or otherwise. In particular, where it can be mounted may simply be a practical decision (for example, it has to be mounted near the point of control or monitoring) or it may be an aesthetic decision (for example, it has to be mounted within an existing enclosure out of sight). The *architecture* of the thing can impact the decision.

Another example of an architectural constraint is a historic building in which wireless IoT is being installed. The building managers are likely going to have strict aesthetic constraints. For example, would you want to visit the great ancient buildings of China, Russia, or Europe only to see them cluttered with little white square IoT devices everywhere? Or would you want to visit Independence Hall in Philadelphia and feel like you've just walked into a modern high-tech business? In both cases, the answer is probably no. Such buildings require careful planning for aesthetics including cable runs, when required, skinning of devices to blend nearly invisibly into the environment, and other methods all while still supporting effective communications for the intended purpose[74].

Budgetary

Budgets are set by the business to assign a value to the IoT system being designed. Sometimes these are firm, not to exceed targets, and in other cases they are general guidelines. When the business creates the overall budget for the project, they are setting a boundary to stay within.

Sometimes a budget is not determined, and the organization will rely on the design engineer's technical expertise to determine the benefit that will be acquired and then the

---

[74] These are just examples. Aesthetic concerns can come up in any IoT project. Commercial buildings, school buildings, enterprise offices, small business, it really doesn't matter. Aesthetic requirements are based on the building management policies of an organization and the desires of the acquirer and stakeholders.

budget that will be provided. Such a scenario will require a kind of rolling design plan where a tentative design may be created without a site visit that can provide a rough order of magnitude (ROM) budget that should be within ±50 percent of the final budget. This goal can usually be achieved because the design professional will make a very close approximation of the number of gateways required, the number of end devices (if they are included in the budget), the number of supporting services required, and so forth. What will change is more about RF in the space than about the other items required.

Some projects used a top-down budget and others use a bottom-up budget. In the top-down approach, you are given a budget and asked to design and deploy a wireless IoT solution within that budget. If the budget is a hard constraint, some features and capabilities may have to be delayed for future upgrades and enhancements. In rare cases, the top-down budget may actually provide more than enough to complete the project.

In the bottom-up approach, you are asked to provide a budget to complete the desired project. In most cases, a ROM budget will be required to gain approval for a more accurate budget later. While this method is more comforting to the supplier, it is also often constrained by contracts that require the solution to be implemented within the budget the supplier specified. For this reason, the supplier must have experts available who can properly determine what is needed and generate an appropriate budget.

In all cases, negotiation of the budget is sometimes required as the details of the project are discovered. If the acquirer adds additional needs during the project, a contract should exist that protects any external suppliers from having to provide those added needs without additional budget. For internal suppliers, the concern is more often schedule than budget, which is discussed later.

Timing / Schedule

Defining the schedule or timing of the project helps set expectations with the stakeholders. If there is a business driver where something must be done quickly, developing a custom solution may not work for the project. If there isn't an off-the-shelf IoT solution and the business needs something quickly, the stakeholders can discuss the needs and may consider a phased approach to delivery. If you are operating as a consultant on the project, having a clear schedule documented ensures alignment with the stakeholders and helps eliminate confusion.

It is also common during either the Design or Deploy phases to encounter the "we want this too" syndrome. It's part of every technology sector and nearly every project. One benefit of an exhaustive requirements engineering process like we've been discussing in this chapter, is that it helps to minimize this issue. However, when something is missed and it is important to the organization, the design or deployment professionals can only seek a method to satisfy the acquirer or stakeholders. Internal suppliers are typically more concerned about the delay in schedule when this occurs (the formal term is scope creep)[75]. External suppliers must be concerned about both the schedule and the budget.

3.10: Industry and Regulatory Compliance

In industrial environments like transportation, oil & gas, manufacturing, mining, utilities, etc. several unique challenges may be presented. When sensors are attached to vehicles in the transportation sector, they need to be able to withstand the range of temperatures, climates, and weather conditions within which the vehicles travel[76]. Oil & gas environments have different environmental conditions from extreme heat in the refineries to salt water on offshore platforms. Understanding the unique characteristics of your environments can ensure the IoT devices are able to hold up over time. The IP Rating Code and NEMA ratings are commonly known examples of these.

For example, a cellular phone rated at IP68 is "dust resistant" and can be "immersed in 1.5 meters of freshwater for up to 30 minutes". For more information on the IP Codes go here: en.wikipedia.org/wiki/IP_Code.

---

[75] Internal suppliers asked to add features or capabilities do not usually "answer for the budget" as much as they "answer for the schedule." However, discussing the impact of both budget and schedule can be important to ensure that the acquirers and stakeholders understand the impact of the added needs they are requesting. These conflicts, though that seems a harsh word in most cases, may require revisiting the requirements to specify new requirements for the addition and ensure they do not conflict with of constrain other requirements. If problems are discovered, they should be communicated with the acquirer so that a decision can be made or that it can be passed through a change management process if one is defined.

[76] Additionally, such wireless IoT devices require mobility support. This means that the IoT network must provide coverage at all locations where the sensors must have communication links. Cellular IoT is often used to provide this, but other solutions like LoRaWAN and Sigfox can work for large coverage areas, and short-range protocols can be used with limited mobility coverage areas exist.

Specific compliance issues in industry include:

- Legal and Regulatory Requirements
- Standards Organizations
- Certification Bodies

Legal and Regulatory Constraints and Requirements

Legal constraints may be general regulatory constraints, or they may be constraints imposed by legal counsel recommendation. That is, a specific and existing regulation may not be the cause of the constraint, but rather the recommendation may be intended to prevent legal liability by going beyond a regulation.

Here are some specific laws or regulatory documents that may apply to your IoT system:

ETSI EN 303 645 Cyber Security for Consumer Internet of Things: Baseline Requirement: This is a European standard for best practices in security for IoT consumer devices. This standard is not likely to apply to the devices in your enterprise or industrial deployments but provides guidelines that can assist in developing best practice baselines for those deployments. It recommends such things as no universal default passwords, vulnerability report management, updates should be secure and timely, secure storage for sensitive security parameters, use cryptography in communications, and minimize exposed attack surfaces.

Radio Equipment Directive (RED): Establishes a regulatory framework for placing radio equipment on the market. A European directive. It specifies guidelines for the implementation of radio equipment such as implementing transmitters that use the radio spectrum efficiently and receivers that have proper reception so as to not require increased transmitter power. It also specifies that radio equipment should be designed to protect personal data and privacy of users. Additionally, among many other things it requires that radio equipment used the Frequency Information System (FIS) of the European Communications Office (ECO).

General Data Protection Regulation (GDPR): As a data privacy regulation, GDPR applies worldwide as any organization that has interactions with individuals in the European Union it likely to have responsibilities related to personal privacy of those individuals. A key element of GDPR is the requirement that organizations consume as little individual personal data as required for a transaction. The information consumed should be adequate, relevant, and limited to that required. The information must be held only when required and not simply for the convenience of the organization. When data is no longer used or when requested by the individual, it must be deleted. Only those requiring access to perform a function should be authorized to access the individual's information and only for the purposes of the function. The complete set of individual rights include the right to be informed, the right of access, the right to rectification, the right to erasure, the right to restrict processing, the right to data portability, the right to object, and rights related to automated decision making and profiling.

Internet of Things Cybersecurity Improvement Act: A US law that establishes minimum security standards for IoT devices owned or controlled by the federal government. It requires the National Institute of Standards and Technology (NIST) to take steps to increase cybersecurity for IoT devices. The developed standards and guidelines should be followed by the federal government and certain organization in partnership with the government. In response, NIST published the IoT Device Cybersecurity Capability Core Baseline[77] and the IoT Non-Technical Supporting Capability Core Baseline[78].

As the CWIDP it is your job to know which of the laws are applicable to your IoT solution and that will govern your project as well as how to adhere to these requirements. The above list is a brief sampling of such regulations. These regulation requirements and others are not tested directly on the CWIDP exam as they vary significantly worldwide.

Additionally occupational health and safety codes and building codes can impact your wireless IoT solution in several ways. They will impact the deployment of the solution, requiring that safety codes are followed while mounting devices, particularly mounting scenarios requiring lifts and ladders. In different parts of the world, such codes are specified and should be adhered to in all aspects related to the IoT project. Issues typically covered include:

- Proper ventilation, temperature, and humidity
- Provision of drinking water
- Adequate lighting

---

[77] www.nist.gov/publications/iot-device-cybersecurity-capability-core-baseline
[78] www.nist.gov/publications/iot-non-technical-supporting-capability-core-baseline

- Environment free from dust, fumes, and other impurities or proper personal protective equipment (PPE) provided
- Prevention of overcrowding with the provision of sufficient workspace to persons
- Prevention of auditory damage through use or PPE
- Implementation of safety procedures related to work-required actions
    - Use of proper safety equipment (harnesses, ladders, lift belts, protective goggles, face shields, face masks, gloves, etc.
    - Working in pairs when required
    - Use of panic devices
    - Secured access to dangerous equipment and products
- Proper training in safety and health procedures

In addition to operational health and safety, building codes are created to assist in ensuring a safe and effective environment for workers. These including codes related to electrical wiring and other cabling, ventilation, lighting, sound management, and construction materials. Specific regions may be more likely to see earthquakes or hurricanes and such realities demand different building codes. The varied building codes influence materials used, which in turn impact RF propagation in the space.

Another area of regulatory restriction, particular to wireless networking, is the regulatory domain in which the network operates. The regulatory domain is defined as a geographically bounded area that is controlled by a set of laws or policies. Currently, governing bodies exist at the city, county, state, and country level within the United States forming a hierarchical regulatory domain system. In other countries, governments exist with similar hierarchies or with a single level of authority at the top level of the country or group of countries. In many cases, these governments have assigned the responsibility of managing communications to a specific organization that is responsible to the government. In the United States, this managing organization is the Federal Communications Commission. In the UK, it is the Office of Communications. In Australia, it is the Australian Communications and Media Authority. The following sections outline four such governing bodies and the roles they play in the wireless networking industry of their respective regulatory domains.

FCC
The Federal Communications Commission (FCC) was born out of the Communications Act of 1934. Charged with the regulation of interstate and international communications

by radio, television, cable, satellite and wire, the FCC has a large body of responsibility. The regulatory domain covered by the FCC includes all 50 of the United States as well as the District of Columbia and other U.S. possessions, like the Virgin Islands and Guam.

OfCom and ETSI

The Office of Communications (OfCom) is charged with ensuring optimal use of the electromagnetic spectrum, for radio communications, within the UK. OfCom provides documentation of and forums for discussion of valid frequency usage in radio communications. The regulations put forth by the OfCom are based on standards developed by the European Telecommunications Standards Institute (ETSI). These two organizations work together in much the same way the FCC and IEEE do in the United States.

MIC and ARIB

In Japan, the Ministry of Internal Affairs and Communications (MIC) is the governing body over radio communications. However, the Association of Radio Industries and Businesses (ARIB) was appointed to manage the efficient utilization of the radio spectrum by the MIC. In the end, ARIB is responsible for regulating which frequencies can be used and such factors as power output levels.

ACMA

The Australian Communications and Media Authority (ACMA) replaced the Australian Communications Authority in July of 2005 as the governing body over the regulatory domain of Australia for radio communications management. Like the FCC in the United States, the ACMA is charged with managing the electromagnetic spectrum to minimize interference. This is done by limiting output power in license-free frequencies, and by requiring licenses in some frequencies.

Outside of the wireless domain, there may be other regulatory bodies and agencies that the IoT system must adhere to. The Occupational Safety and Health Administration (OSHA) in the United States is an example of one such organization. Identifying agencies that will place requirements on your specific system is in many cases a joint effort and consulting legal council is generally a good idea if you are building a system from the ground up.

Standards Organizations

Standards organizations define the operational capabilities and functions of devices, protocols, applications, and security components. Several standards organizations are international, and others are national. Many national standards organizations adopt standards of other nations or international standards to allow for compatibility among electronics devices. Some organizations operate in a general category, such as the IEEE (electronics engineering) and IETF (Internet protocols), while others exist for a specific protocol or technology, such as the Bluetooth Special Interest Group (SIG) or the LoRa Alliance. You will not be tested on the specific of the operational mandate of any such organization but should be aware of the role they place in ensuring consistency on function and interoperability among devices through compliance with the standards. A few organizations related to wireless IoT are listed below:

- Institute of Electrical and Electronics Engineers (IEEE)
- Internet Engineering Task Force (IETF)
- International Organization for Standardization (ISO)
- International Electrotechnical Commission (IEC)
- International Telecommunications Union (ITU)
- American National Standards Institute (ANSI)
- LoRa Alliance
- Connectivity Standards Alliance (CSA) (formerly Zigbee Alliance)
- Bluetooth SIG
- oneM2M
- European Telecommunication Standards Institute (ETSI)
- Australian Communications and Media Authority
- IoT World Alliance
- Industry IoT Consortium

Certification Bodies

Certification bodies certify devices and/or systems to be in compliance with guidelines that ensure interoperability and functionality. Some of the previously mentioned standards organizations also perform device certification. For example, the LoRa Alliance certifies products for use on LoRaWAN networks[79]. Additional organizations include but are not limited to:

- Wi-Fi Alliance
- LoRa Alliance
- Connectivity Standards Alliance
- Bluetooth SIG
- Microsoft Azure
- AT&T

3.11: Quality Requirements Specification

The 29148-2018 standard, explained in the preceding sections, references the IEEE 15288-2015 standard, which in turn references the ISO/IEC 25030-2019 standard in relation to quality as a requirement in systems. Quality is an important requirement in wireless IoT solutions from several perspectives and this section will explain the 25030-2019 standard[80] and apply it to an IoT solution. Figure 3.6 provides a high-level overview of the processes involved in establishing quality requirements based on these standards[81]. The development of quality requirements begins during the business requirements and stakeholder needs and requirements phases of requirements engineering and continues into the system or technical requirements phase covered in the next chapter.

---

[79] lora-alliance.org/lorawan-certification/
[80] The ISO/IEC 25030-2019 standard is part of the 25000 series of standards, which are focused on System and Software Quality Requirements and Evaluation (SQuaRE). ISO/IEC 25001-2014 provides the foundation, 25010 defines the Quality in Use Model and Product Quality Model, while 25012 defines the Data Quality Model.
[81] It is highly recommended, as an IoT solutions design professional, that you become intimately aware of the requirements standards available from the IEEE and ISO/IEC. They are based on decades of experience and bring tremendous value to the knowledge base of any network or systems engineer. We are providing sufficient information here to understand their implementation and use in wireless IoT scenarios but exploring them in-depth can result in even greater requirements engineering skills and successes in your career.

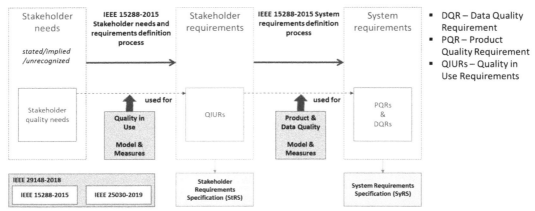

Figure 3.6: ISO 25030-2019 Relation to Other Requirements Engineering Standards by the IEEE, ISO, and IEC (Reproduced from ISO/IEC 25030-2019)

Some terms in Figure 3.6 require an explanation:

- **Quality in Use Model:** A quality model defining five characteristics related to the outcomes of interaction with a system: effectiveness, efficiency, satisfaction, freedom from risk, and context coverage.
- **Product Quality Model:** A quality that categorizes system quality properties into eight characteristics: functional suitability, performance efficiency, compatibility, usability, reliability, security, maintainability, and portability.
- **Data Quality Model:** A quality model that defines fifteen characteristics for data quality requirements. Data quality requirements may be derived from QIURs and PQRs. The larger the generated data set, the more important data quality requirements become. A massive data set with no context, with no cleansing, or with no value to the user may be as useless as no data at all.
- **QIURs:** Quality in Use Requirements are stakeholder-level quality requirements.
- **PQRs:** Product Quality Requirements are system-level quality requirements.
- **DQRs:** Data Quality Requirements are quality requirements for data.

From Figure 3.6, you can determine that the stakeholder needs are used to derive stakeholder requirements. These requirements are based on the Quality in Use Model and result in Quality in Use Requirements (QIURs). From the QIURs we derive the Product Quality Requirements (PQRs based on the Product Quality Model) and the Data Quality Requirements (DQRs based on the Data Quality Model). The PQRs and DQRs are of most importance to us and, in fact, the DQRs are defined in a completely different standard from the QIURs and PQRs (ISO/IEC 25012 defines the data quality model and ISO/IEC 25024 defines the measurement of data quality).

To further illustrate how quality requirements are derived from processes with users and stakeholders as the source of needs leading to requirements, and from other information and communication technologies (ICT) systems as constraints, consider Figure 3.7. This figure illustrates the derivation of quality requirements from different sources at different levels. Users and other stakeholders, such as regulators, managers, and industry standards groups, lead to QIURs or stakeholder requirements. Then, within the information system solution, multiple ICTs may exist. For example, the wired network infrastructure could be considered an ICT and the individual services running on that network, such as DHCP, NTP, and Internet access, could be considered ICTs; however, they are not the target ICT under the scope of requirements development for the wireless IoT network.

If we take the wireless IoT network to be the inner-most ICT on the left side of the image, PQRs (our primary concern with an IoT network) will be derived from the QIURs and the environment in which the network will operate. Our greatest concern will be the requirements for the network in the context of the existing systems, which are in the context of the user-base, which provided the QIURs. You can begin to see the hierarchy of derivation. At the same time, the non-target ICTs, for example, the wired network and network services previously mentioned, certainly give requirements as constraints to the wireless IoT network[82].

For example, if we cannot upgrade the wired network, the IoT network will be limited to the current capabilities of the wired network. If we cannot add new services, the IoT network will be limited to the capabilities of the existing services. At the same time, we should not assume that the wired network cannot be upgraded or that the services cannot be upgraded. Knowing what is available on the wired network, in relation to

---

[82] At this point, I hope you're beginning to see the value of the standards-based requirements engineering processes that we began exploring in this chapter and continue to explore in the next. They provide an excellent model and, possibly of greater importance, they provide a strong argument to go to management and plead for effective requirements engineering in the design process. The first time or two that you use these models in a design project, they may seem overwhelming, but just like mastery of the IoT protocols, they quickly become second nature.

supporting services for the IoT network is important both for recommendation upgrades and for designing within constraints.

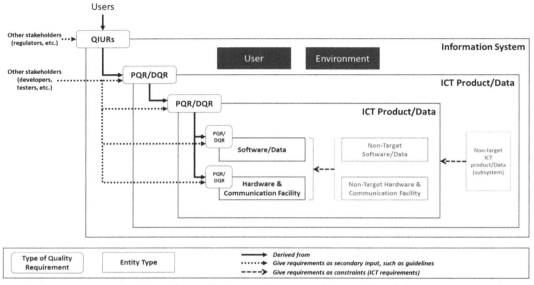

Figure 3.7: Derivation of Quality Requirements (Reproduced from 25030-2019)

Now, I want to be clear on one thing. In the previous explanations I stated that the PQRs were most important to the wireless IoT network and, indeed, they are. But the IoT solution encompasses the entire IoT architecture, which does include the data element. Therefore, DQRs are very important to the overall IoT solution. The DQRs will specify the structure, format, accuracy, and contents of the required data, among other elements. One important quality requirement related to data in IoT solutions is traceability. Traceability is the degree to which data has attributes that provide an audit trail of access to the data and of any changes made to the data in a context of use[83]. In an IoT solution, this also demands that the data can be traced back to the IoT end device of origination. For example, it's not much use to know that the temperature is 101 degrees Fahrenheit somewhere. I need to know the specific source device so that I can identify the location where that temperature was detected. While this extends the concept of traceability beyond that in the ISO/IEC 25012 standard, it is an important extension for IoT and

---

[83] ISO/IEC 25012-2008 Data Quality Model Clause 5.3.2.6 Traceability

ensures both the completeness and credibility data quality characteristics that are equally important.

As stated previously, the Quality in Use Model defines five characteristics. These characteristics are represented in Figure 3.8.

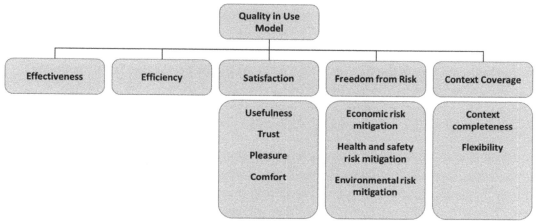

Figure 3.8: Quality in Use Model with Characteristics and Sub-Characteristics

Effectiveness is the accuracy and completeness with which users can achieve specific goals with the system. In a wireless IoT network, this will be most impacted by connectivity and consistent connectivity. The focus is on completeness through the ability to maintain a stable connection throughout the activity performed or when an activity is required. In some cases, this stakeholder need may demand that the device in use can maintain connectivity while the device moves throughout the facility or the coverage space. If the connection is lost for a significant amount of time, the application may reset causing a loss of previously performed actions, or a state of incompleteness. A stakeholder requirement statement coming out of this quality characteristic might be: *The wireless IoT network shall provide connectivity in the warehouse as forklifts and other machinery move from place to place with transfer actions that do not result in the loss of connectivity for the application.*

Efficiency is a quality characteristic that encompasses the resources expended in relation to the accuracy and completeness with which users or devices achieve specific goals with the system. The focus is on accomplishing effectiveness with only the required resources and while avoiding overengineering. A stakeholder requirement statement coming out of this quality characteristic might be: *The wireless IoT network shall be implemented using the required number of network connection points to achieve effectiveness and satisfaction and no more.*

Satisfaction is the first quality characteristic in the model to have sub-characteristics. Satisfaction is the degree to which user needs are satisfied when a product or system is used in a specified context of use. It is a factor of usefulness, trust, pleasure, and comfort. *Usefulness* is a measure of the perceived ability to achieve goals using the system. *Trust* is a measure of the degree to which the user can expect the system to function as intended. *Pleasure* is a measure of the degree to which a used fulfills personal needs through the use of the system, including acquiring new skills and knowledge, communicating personal identity, and provoking pleasant memories. Finally, *comfort* is a measure of the degree to which the user is satisfied with physical comfort when using the system.

As an illustration of the last sub-characteristic, think of the commercials and movies where people are standing on chairs, rooftops, or other uncomfortable locations in an attempt to gain a signal for their cellphones. This is an example of poor quality in relation to the comfort sub-characteristics. A stakeholder requirement statement coming out of this quality characteristic (satisfaction) might be: *The wireless IoT network shall function in all areas of the facility such that devices or users are never required to change locations to gain access and that function shall provide stable and consistent connections for the achievement of lengthy processes which require state endurance over an extended period*[84].

Freedom from risk is a quality characteristic that defines the degree to which the product or system mitigates the potential risk to economic status, human life, health, or the environment. It includes three sub-characteristics: economic, health and safety, and environment risk mitigation. A stakeholder requirement statement coming out of this

---

[84] This is a good place for a point of clarification. At this stage, you are not defining system requirements. You are defining stakeholder requirements. From the stakeholder requirements you will derive system requirements. As you design the IoT solution, the system requirements specification will guide your design. As long as they were developed with proper links to stakeholder/user requirements and business requirements, you should not be required to recursively evaluate the source of the system requirements during the design process. However, when the system requirements are unclear or you need more information, going back to the business and stakeholder requirements can prove beneficial. Certainly, in the validation phase, you will validate that the business and stakeholder requirements were met through the system requirements and solution design and deployment.

quality characteristic might be: *The wireless IoT network shall be implemented within {regulatory agency} guidelines to avoid exposure to harmful RF energy and shall be powered through energy harvesting solutions that minimize energy consumption from the power grid.* Another example statement might be: *The wireless IoT network shall be implement with proper security to protect against the theft of personal privacy information or corporate intellectual property that could cause financial harm to the individual or organization.*

Context coverage is the final quality characteristic in the model and is a measure of the degree to which the system can be used with the other four characteristics in the context or environment in which it will operate. This characteristic includes two sub-characteristics: context completeness and flexibility. *Context completeness* is a measure of the ability of the system to be used with effectiveness, efficiency, satisfaction, and freedom from risk in all specified contexts of use. Stated differently, you could say that the system can be used in all requirement areas while meeting the effectiveness, efficiency, satisfaction, and freedom from risk requirements in those area. *Flexibility* is a measure of the degree to which the system can be used in contexts beyond those initially specified in the requirements. That is, how challenging will it be to adapt the selected wireless IoT solution to new requirement area definitions or completely new locations?

As you can see, the Quality in Use Model surfaces many important considerations in the design of a wireless IoT network and solution. Such quality requirements should be considered in the requirements engineering process and for each requirement area, when needed.

The Product Quality Model is used to define quality requirements at the system level. It is well-documented in the ISO/IEC 25010-2011 standard and brief definitions of the characteristics and sub-characteristics follow. It categorizes the product quality requirements into eight characteristics: functional suitability, performance efficiency, compatibility, usability, reliability, security, maintainability, and portability[85]. The product quality model can be applied to just a software product, or to a computer system that includes software, as most of the sub-characteristics are relevant to both software and systems. Figure 3.9 illustrates the Product Quality Model used for system quality requirements.

---

[85] ISO/IEC 2010-2011 System and Software Quality Models Clause 3.3 Product Quality Model

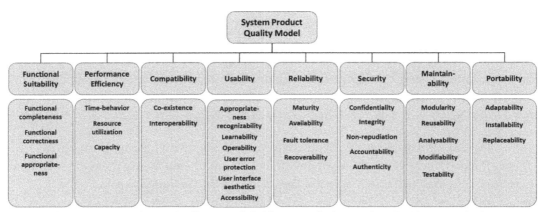

Figure 3.9: Product Quality Model Characteristics and Sub-Characteristics

The characteristics and sub-characteristics are defined as[86] (those most important to wireless IoT solution requirements are indicated in color, though all may apply in some contexts):

- **Functional Suitability:** degree to which a system provides functions that meet stated and implied needs when used under specified conditions
    - *Functional completeness:* degree to which the set of functions covers all the specified tasks and user objectives
    - *Functional correctness:* degree to which a system provides the correct results with the needed degree of precision
    - *Functional appropriateness:* degree to which the functions facilitate the accomplishment of specified tasks and objectives
- **Performance Efficiency:** performance relative to the amount of resources used under stated conditions (spectrum can be considered as a resource in wireless networking system)
    - *Time-behavior:* degree to which the response and processing times and throughput rates of a system, when performing its functions, meet requirements
    - *Resource utilization:* degree to which the amounts and types of resources used by a system, when performing its functions, meets requirements (consider spectrum for Wi-Fi)

---

[86] ISO/IEC 25010-2011, clause 4.2, Product Quality Model

- *Capacity:* degree to which the maximum limits of a system parameter meet requirements (parameters can include number of concurrent users, communication bandwidth, throughput of transactions, etc.)
- **Compatibility:** degree to which a system can exchange information with other products, systems, or components, and/or perform its required functions, while sharing the same hardware or software (or spectrum) environment
    - *Co-existence:* degree to which a system can perform its required functions efficiently while sharing a common environment and resources with other products (or systems), without detrimental impact on any other product (or system)
    - *Interoperability:* degree to which two or more systems can exchange information and use the information that has been exchanged (in the context of Wi-Fi: roaming capabilities)
- **Usability:** degree to which a system can be used by specified users to achieve specified goals with effectiveness, efficiency, and satisfaction in a specified context of use
    - *Appropriateness recognizability:* degree to which users can recognize whether a system is appropriate for their needs
    - *Learnability:* degree to which a system can be used by specified users to achieve specified goals of learning to use the system with effectiveness, efficiency, freedom from risk, and satisfaction in a specified contest of use
    - *Operability:* degree to which a system has attributes that make it easy to operate and control
    - *User error protection:* degree to which a system protects users against making errors
    - *User interface aesthetics:* degree to which a user interface enables pleasing and satisfying interaction for the user
    - *Accessibility:* degree to which a system can be used by people with the widest range of characteristics and capabilities to achieve a specified goal in a specified context of use
- **Reliability:** degree to which a system performs specified functions under specified conditions for a specified period of time
    - *Maturity*: degree to which a system meets needs for reliability under normal operations

- *Availability*: degree to which a system is operational and accessible when required for use
- *Fault tolerance*: degree to which a system operates as intended despite the presence of hardware or software faults
- *Recoverability*: degree to which, in the event of an interruption or a failure, a system can re-establish the desired state of the system
- **Security:** degree to which a system protects information and data so that persons or other systems have the degree of data access appropriate to their types and levels of authorization
    - *Confidentiality*: degree to which a system ensures that data are accessible only to those authorized to have access
    - *Integrity*: degree to which a system prevents unauthorized access to, or modification of, computer programs or data
    - *Non-repudiation*: degree to which actions or events can be proven to have taken place, so that the events or actions cannot be repudiated later
    - *Accountability*: degree to which the actions of an entity can be traced uniquely to the entity
    - *Authenticity*: degree to which the identity of a subject or resource can be proved to be the one claimed
- **Maintainability:** degree of effectiveness and efficiency with which a system can be modified by the intended maintainers
    - *Modularity*: degree to which a system is composed of discrete components such that a change to one component has minimal impact on other components
    - *Reusability*: degree to which an asset can be used in more than on system, or in building other assets
    - *Analysability*: degree of effectiveness and efficiency with which it is possible to assess the impact on a system of an intended change to one or more of its parts, or to diagnose a system for deficiencies or causes of failures, or to identify parts to be modified (monitoring in WLANs)
    - *Modifiability*: degree to which a system can be effectively and efficiently modified without introducing defects or degrading existing product quality

- *Testability*: degree of effectiveness and efficiency with which test criteria can be established for a system and tests can be performed to determine whether those criteria have been met
- **Portability**: degree of effectiveness and efficiency with which a system can be transferred from one hardware, software or other operational or usage environment to another (from cloud management to on-premises, from autonomous to controller-based, etc.)
  - *Adaptability*: degree to which a system can effectively and efficiently be adapted for different or evolving hardware or other operational usage environments (adding 802.11ax, incorporating 6 GHz, etc.)
  - *Installability*: degree of effectiveness and efficiency with which a system can be successfully installed and/or uninstalled in a specified environment (short-term WLANs for mobile operations, etc.)
  - *Replaceability*: degree to which a system can replace another system for the same purpose in the same environment (WLAN upgrades)

Finally, the Data Quality Model defines fifteen characteristics of data that should be met and it is defined in ISO/IEC 25012-2008 (renewed in 2019). Figure 3.10 lists the 15 characteristics of data quality. Inherent data qualities are those that have the intrinsic potential to meet the needs (stated and implied) when data is used in specific conditions. That is, the data can meet the requirement without the aid of a specific system. System dependent data qualities are those that demand specific capabilities in the system to be provided. Some qualities are achieved through a combination of the intrinsic potential and that offered by the system (the middle listing in Figure 3.10).

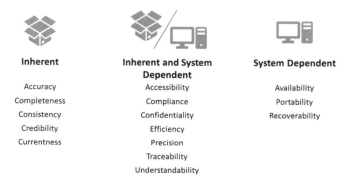

Figure 3.10: Data Quality Model Characteristics

As you can see, this standard set provides an exhaustive set of characteristics and sub-characteristics to consider for system quality requirements. The wireless IoT solution you are planning is the product, and the system quality requirements define the quality-related capabilities that should exist in the system.

3.12: Writing Requirements: A Case Study

This case study will use the California requirement for employees to have a means of indicating trouble when in a guestroom in the hospitality industry. Let's start with breaking apart California AB1761 and turning it into requirements, here is the first clause and we will utilize this to create proper requirements.

> Provide employees working alone in a guestroom with a panic button, free of charge. The employee may use the panic button, and cease work, if the employee reasonably believes there is an ongoing crime, harassment, or other emergency happening in the employee's presence. The hotel employer shall develop an appropriate protocol, including any necessary training, for how staff, security, and management shall respond when a panic button is activated. The protocol shall be calculated to ensure an immediate on-scene response to the greatest extent possible.

The bill goes on to clarify what is meant by Employee, Hotel employer, and Panic button.

> (1) "Employee" means an individual who, in any particular workweek, performs at least two hours of work for a hotel employer. "Employee" also includes a subcontracted worker.

> (2) "Hotel employer" means a person, including a corporate officer or executive, who directly or indirectly, including through the services of a temporary staffing service or agency, employs or exercises control over the wages, hours, or working conditions of employees at a hotel, motel, bed and breakfast inn, or similar transient lodging establishment as defined in Section 1865 of the Civil Code and includes any contracted, leased, or sublet premises connected to or operated in conjunction with the purpose of the lodging establishment.

(3) "Panic button" means an emergency contact device that an employee can use to summon immediate on-scene assistance from another employee, security personnel, or representative of the hotel employer.

The first sentence here breaks up as follows with an implied subject, modified to form the construct.

Hotel owner [subject] shall provide [action] employees working alone in a guest room [constraint of action] with a panic button (emergency contact device) [object], free of charge [condition].

The employee [subject] may use the panic button, and cease work, [action] if the employee reasonably believes there is an ongoing crime, harassment, or other emergency happening in the employee's presence [constraint of action].

The hotel employer [subject] shall develop an appropriate protocol, including any necessary training, for how staff, security, and management shall respond [action] when a panic button is activated [constraint of action].

When we break down the above, it fails on the "Is this singular" requirement. Another way to state this is to break this into multiple requirements, like this:

The hotel employer [subject] shall develop an appropriate protocol for how staff, security, and management shall respond [action] when a panic button is activated [constraint of action].

The hotel employer [subject] shall develop any necessary training for how staff, security, and management shall respond [action] when a panic button is activated [constraint of action].

The hotel employer [subject] shall deliver any training for how staff, security, and management shall respond [action] when a panic button is activated [constraint of action].

The protocol [subject] shall be calculated to ensure an immediate on-scene response [action] to the greatest extent possible [constraint of action].

In the last sentence of requirement one, the constraint of action is not verifiable because of the word "possible". This could be reworded; however, for the purposes of these business requirements we will leave this alone.

If we take these requirements, we can expand upon them easily. Start with this checklist:

- What is the primary use case / justification for the project?
- Where will the network operate?
- Are there areas to specifically exclude from the network?
- Who will use the system?
- What interfaces will be part of the system?
- What data will be collected from the system?
- What is the impact if the system is unavailable?

The primary use case and justification are covered within the bill itself. Given that it is required by law in some areas, it is justified because it is required.

The next question is, Where will the network operate?

There are two items within the requirements that address where the panic button will operate and, therefore, where the network must be available. The first is the clause "employees working alone in a guest room" and the second is "in the employee's presence." Since this law is covering the hotels, it is safe to state this as a requirement in this fashion. *The panic button shall have service within the guest rooms and throughout the hotel.*

Are there areas to specifically exclude from the network? The bill doesn't specify this, so there is no need to create a requirement around exclusion areas.

Who will use the system? It is clear that hotel staff, security, and management will use the system. But how will they interface with the system? The bill doesn't define anything more than "develop an appropriate protocol for how staff, security, and management shall respond." This is fine for the purposes of the bill; however, to implement the IoT system this will need to be expanded upon.

The employees in the room will have a panic button, which has already been defined as their interface. Security and hotel management must have another means of notification to respond, and a requirement needs to be written around this. That said, the response begs the question of the location of the employee. How can you respond without knowing the location that the panic button has been activated? A requirement for this could be the following:

Upon activation of the panic button an alert shall immediately provide the room number or location on a map to hotel security and management[87].

The interface that hotel security and management will use is also not defined within the bill. It allows for the creation of the interface as part of "develop an appropriate protocol," as this will hopefully integrate into what the hotel has today. It could be a text message alert. It may be an alert on the security monitors or both of these or something else that works for the staff. A requirement should be written around this.

What data will be collected from the system itself? This needs to be answered from the business perspective, but the bill doesn't address this. The business owner may want to add some requirements here around reporting how often the panic buttons are used, where they are used, logging, tracking when panic buttons are offline, battery levels, etc.

The last question is about the impact if the panic button is unavailable, having the system unavailable would violate the law and can be written into the requirements as such. We will leave this impact to the legal minds and assume the network is of critical importance for life safety issues and leave this alone from the requirements standpoint.

Now that the first pass at the business requirements is complete, let's look at the list of constraints and make sure we have requirements around these complete as well. As a refresher here is a quick checklist of possible constraints to create requirements around:

- What is the schedule/deadline for the proposed system?
- Are there architectural or aesthetic considerations?
- Is there a budget limit for the project?
- What Regulatory Compliance domains will the system operate within?
- Are there industry specific concerns?
- Are there specific safety requirements or compliance concerns?

Here are a few samples of constraints utilizing the above checklist:

- The panic button solution shall be fully operational, verified, and tested prior to Jan 1, 2021.

---

[87] In fact, you may desire to know the location of the employee and of the panic button. The former will tell you where the employee is currently located (have they escaped the room or been taken from the room since pressing the panic button) and the latter will tell you the room in which the button was pressed.

- Elements of the panic button solution shall not be visible or approved by the general manager of the hotel prior to deployment.
- The panic button solution shall cost less than $100 per guest room to purchase, deploy, and test.
- All wireless components of the IoT system will be compliant to operate within the FCC domain.
- The panic button solution shall adhere to United States and California laws.

3.13: Chapter Summary

In this chapter, you learned how to properly engineer requirements with a focus on business requirements. The proper construct of a requirements statement includes a condition, subject, action, object, and constraint of action. Which keywords indicate a mandatory vs non-mandatory vs suggestion for a requirement, as well as, keywords to avoid within requirements. After each requirement is constructed and validated you also discover how to track those requirements as part of the IoT project itself.

To summarize requirements engineering, before looking more deeply at the technical requirements definition process in chapter 4, considered Figure 3.10. We start with the inputs of acquirer/business and stakeholder needs to translate them to business and stakeholder requirements. Then we create and validate technical requirements of the system or solution. Finally, we design the solution based on these requirements. However, at any time in the design effort, conflicts and issues can occur that require iteration or recursion back into the requirements definition process.

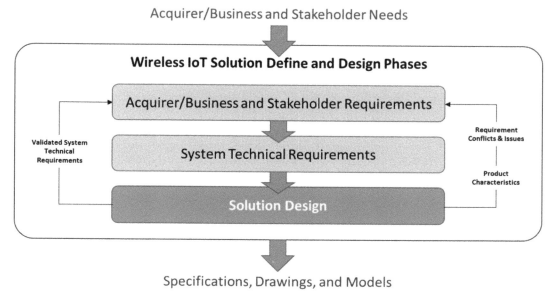

Figure 3.10: Wireless IoT Solution Define and Design Process[88]

---

[88] Enhanced from EIA-632 Processes for Engineering Systems

Additionally, we have created a sample spreadsheet that can be found here: bit.ly/3kInxVu

This sample spreadsheet was created from the 29148-2018 guidelines and is free for use courtesy of CWNP. It includes reminders for proper language use and checklists to ensure requirements statements are well-formed.

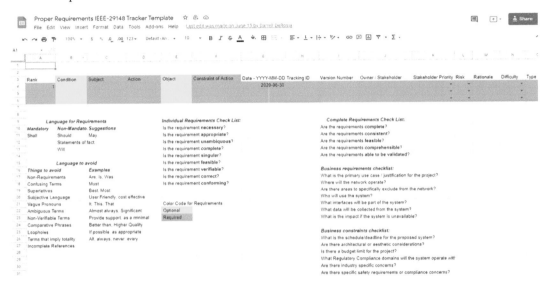

3.14: Review Questions

1. What IEEE standard defines the requirements engineering processes?

    a. IEEE 29148-2018

    b. IEEE 802.11

    c. IEEE 802.15.4

    d. IEEE 802.15.6

2. Which one of the following comes from the acquirer domain in interdisciplinary requirements engineering?

    a. Technical Professionals

    b. IoT Solutions

    c. Stakeholders

    d. Technical Expertise

3. What is a minimum set of stakeholders per IEEE 29148-2018?

    a. Acquirers and Suppliers

    b. Acquirers and Users

    c. Suppliers and Users

    d. None of these

4. Which of the following are characteristics of individual requirements based on IEEE 29148-2018? (Choose all that apply.)

    a. Necessary

    b. Feasible

    c. Unambiguous

    d. Comprehensible

5. True or False: A requirement statement is singular if it states a single capability, characteristic, constraint, or quality factor.

    a. True

    b. False

6. Which one of the following is a TBx designation according to IEEE 29148-2018?

    a. To Be Defined

    b. To Be Secured

    c. To Be Renewed

    d. To Be Synchronized

7. Which one of the following represents the construct recommended by IEEE 29148-2018 for a requirement statement?

    a. [Condition][Action][Subject][Object][Constraint of Action]

    b. [Condition][Subject][Action][Object][Constraint of Action]

    c. [Condition][Action][Subject][Object]

    d. [Condition][Object][Subject][Action][Constraint of Action]

8. What term is used to describe a requirement that defines what a system must allow a user to do?

    a. Non-Functional

    b. Functional

    c. Human Factors

    d. Quality

9. Which one of the following terms should be avoided in requirement statements?

    a. Shall

    b. Should

    c. Most

    d. None of these

10. What requirement statement attribute is used to perform requirements tracing throughout a project life cycle?

    a. Owner

    b. Identification

    c. Stakeholder Priority

    d. Type

3.15: Review Answers

1. **The correct answer is A.** IEEE 29148-2018 defines the requirements engineering processes. Specifics of certain processes are also defined in other standards.

2. **The correct answer is C.** Stakeholders come from the acquirer domain. The other listed items come from the supplier domain.

3. **The correct answer is B.** Suppliers are not necessarily stakeholders, so the correct answer is acquirers and users.

4. **The correct answers are A, B,** and **C.** Comprehensible is a characteristics of a requirements set and not of individual requirements or requirement statements.

5. **The correct answer is A. True.** This statement accurately describes a singular requirement statement.

6. **The correct answer is D.** TBx designations include To Be Defined (TBD), To Be Specified (TBS) and To Be Resolved (TBR).

7. **The correct answer is B.** [Condition][Action][Subject][Object][Constraint of Action] is the construct recommended.

8. **The correct answer is B.** Functional requirements define what a user of a system should be able to do or what tasks they must perform. Non-functional requirements define the 'ilities' such as reliability, survivability, and maintainability.

9. **The correct answer is C.** Most is an imprecise term and should be avoided. Shall, should, and may are commonly used with intended meanings defined within the requirements set.

10. **The correct answer is B.** The identification attribute can be used to track or trace a requirement throughout the life cycle. It is typically a numeric, possibly hierarchical numeric identifier.

Chapter 4: Defining Technical Requirements

Objectives Covered:

2.2 Gather technical requirements and constraints

In the previous chapter, we reviewed the business and stakeholder requirements and how to create good requirements in general. Now that we understand creating requirements in more detail, we have a firm understanding of what we need to do. Next, we can focus on how to deliver the solution that meets the business requirements and where the IoT systems will be deployed. In this context of technical requirements, *where* relates to the location where a technical function is performed within the system. For example, converting data from one format to another is performed in an ETL solution. *Where* is within the ETL solution that exists or will be created in your project. *Where* can also include the location in physical space where functions must be performed, such as where IoT devices must be located to perform sensing or where the human must be able to view information in a dashboard.

This chapter, then, provides details on creating technical requirements. We look at the process from 29148-2018 for converting business and stakeholder requirements into technical requirement and then we explore the various components and characteristics of a wireless IoT solution for which technical requirements should be specified. First, let's investigate the meaning of technical requirements.

4.1: Technical Requirements Defined

Technical requirements, also called system or software requirements, can be broadly grouped into two categories, functional and non-functional requirements[89]. Functional requirements are ones that provide functionality to the IoT system while non-functional requirements may enhance functionality but not specifically provide a function themselves. Stated differently, a functional requirement defines what a system can technically do, and a non-functional requirement defines the qualities of the action such as security, performance, etc. You could also simplify it to this: functional equals what the system does; non-functional equals how the system does it, if *how* is understood to

---

[89] The ISO/IEC 25030-2019 standard defines functional requirements as *requirements that specify a function that a system or system component shall perform*. Therefore, that which is *non-functional* would be that requirement which does not specify a function performed, but rather some quality of the system or function. For this reason, quality requirements and non-functional requirements are often used synonymously.

mean the characteristics of the action and not the steps of the action. Functional requirements define the steps of the action and non-functional requirements define the characteristics surrounding that those steps.

Here's a simple example: The functional requirement of a vehicle may be that it must move forward when in the forward gear and the gas pedal is pressed. The non-functional requirement may be that it must reach 60 MPH in seven seconds. Any car can meet the functional requirement, meaning that any care can indeed move forward when the gas pedal is pressed; however, it takes a special quality of car to reach 60 MPH in seven seconds.

From this example, you can see that a system can work and do all the tasks required of it, but that doesn't mean that it will meet the user's expectations. If it is too slow, often unavailable, lacking security, or any other desired quality the users desire, it will fail to meet expectations.

For the CWIDP certification here is a high-level view of which requirements are functional vs. non-functional, though not an exhaustive list.

- Functional requirements include:
  - System capabilities
  - Interfaces
    - Connectivity
    - User interfaces
    - System interfaces
  - System operations
    - ETL processes
    - Programming scripts
- Non-Functional requirements (often categorized as quality requirements) include:
  - Usability
  - Security
  - Performance
  - Reliability
  - Scalability
  - Coexistence

It can help to diagram the flow of the end-to-end system you envision before putting the requirements in writing. Figure 4.1 is a sample diagram of one such system. This is also referred to as an architectural diagram. In some cases, a basic architectural diagram is designed before the technical requirements are specified to assist in the process. It does not necessarily list every technology or device required to implement a functional solution, but it provides an overview of what is required. For example, the sensor DB may be a three-server cluster running Hadoop for big data processing or it may be a single MongoDB database. The technical specifications of the design will specify this based on the technical or system requirements. It is not uncommon to develop a preliminary architecture model that is then enhanced throughout the requirements engineering process. The result is an architecture that is mostly complete at the end of requirements engineering.

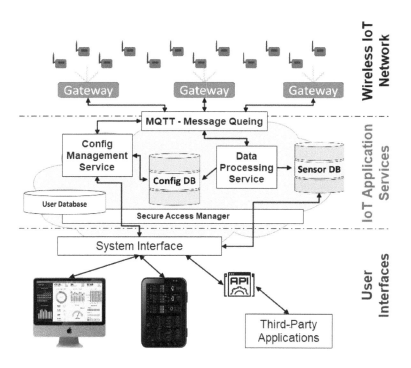

Figure 4.1: Basic IoT Architectural Diagram

Both functional and non-functional requirements should be captured to create the technical requirements for the IoT Project. Remember, the goal of the technical requirements is to enable the solution to meet the business and stakeholder requirements. As a result, each technical requirement should trace back to a specific business or stakeholder requirement. If there is not a means to provide a technical solution within the constraints provided by the business, there should be discussion between the stakeholders regarding options and changes to those requirements.

4.2: The IEEE 29148-2018 Technical Requirements Process

To understand the standard requirements engineering processes for system or technical requirements, you must first understand the conceptual nature of a system. Then, you can understand the processes used to define the technical requirements for a system or several systems. We will start with system concepts.

System Concepts

A *system* is a combination of interacting elements organized to achieve one or more stated purposes[90]. The intent of a system is typically phrased such that a descriptor is used with the word system. For example, an IoT system, a database system, or a big data system are each specific types of systems. The complete system is inclusive of all the equipment, facilities, material, software, firmware, technical documentation, services, and personnel required for operations and support.

A *system element* is a member of the set of elements that constitute the system[91]. That is, the list stated in the previous paragraph is a list of system elements (equipment, facilities, material, etc.). If S represents the set of system elements and SE represents a system element, you can say[92] that $SE \in S$. Each system element can be implemented independently to perform a function, provide a service, or meet some other requirement. However, it is part of a system because it interfaces with other elements within the system. Additionally, it may be true that system one (S1) and system two (S2) both use a system element making the following true: $SE \in S1 \text{ and } S2$. Figure 4.2 illustrates the

---

[90] IEEE 29148-2018 clause 4.1.46
[91] IEEE 29148-2018 clause 4.1.47
[92] Read *SE is a member of the set S* or *SE is in S* or *SE is an element of set S*.

concept of a system and system elements. Figure 4.3 illustrates the concept with the example of an IoT system. The *system of interest* is the system under design and development. It is the focus of the requirements engineering process, which may demand requirements engineering for multiple systems.

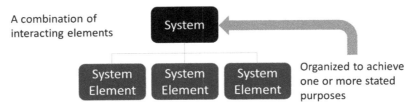

Figure 4.2: A System and its Elements

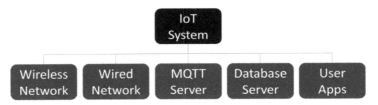

Figure 4.3: An IoT System and its Elements

In some cases, systems shown in Figure 4.3 may be considered *enabling systems* or supporting services. For example, the wired network may already exist and our IoT system is simply going to use it; therefore, it is an enabling system. Figure 4.4 illustrates the concept of enabling systems. The operational systems are those that may require interfaces with the system of interest, but do not necessarily provide functionality to it.

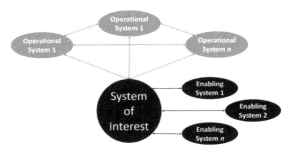

Figure 4.4: Enabling Systems

A System of Systems (SoS) is a *set of systems that integrate or interoperate to provide a unique capability that non of the constituent systems can accomplish on its* own. For example, IoT end devices and gateways, alone, do not provide a valuable IoT solution, though they are generating data. We need systems to process the data, store the data, analyze the data, act on the data and so forth. Figure 4.5 illustrates the general concept of an SoS and Figure 4.6 illustrates a potential (and partial) IoT SoS.

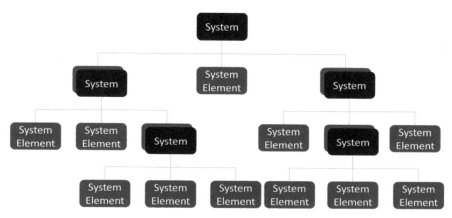

Figure 4.5: System of Systems Concept

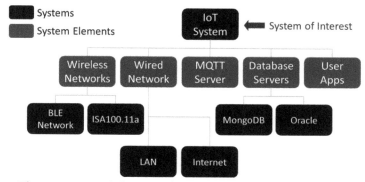

Figure 4.6: Partial Example of an IoT System of Systems

At times, the top-level system will be the system of interest. The one that you are focused on designing. At other times, using Figure 4.6 as a reference, instead of the entire IoT system, you will be called in to design just a portion of the overall system, such as the BLE network. In such cases, for you, the system of interest will be the wireless BLE network, and the other related systems will be enabling systems. With system of systems thinking, as opposed to simply systems thinking, you must design the system in the context of the greater whole. Figure 4.7 illustrates this concept. Here, you can see that the BLE network system is the system of interest, but it must exist in the context of the others. For example, the ISA100.11a network must coexist in the same frequency band as the BLE network and the designer must consider this. Additionally, the BLE network may consume services from the Database Servers or the Wired Network to perform its full function.

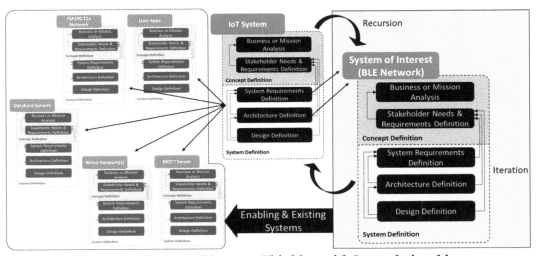

Figure 4.7: System of Systems Thinking with Interrelationships

The requirements gathering and defining process is cyclical or iterative in nature, meaning that the process itself will sometimes go in a circle. This is normal and to be expected because business requirements are created without insight into the technical nature of implementation of those requirements. Figure 4.7 shows both iteration (within a system) and recursion (between systems). As more information is collected and gathered about the overall (business and technical) requirements things may be found within the requirements which are either not feasible or that cost more than allocated to make them happen. In these instances, discussions will need to occur with the owners of the business and technical requirements to discuss what options are available, what makes sense, and ultimately decide how to move forward with the project.

Technical Requirements Engineering Process

The technical requirements for a system, as previously stated, are also called system requirements. The technical requirements should include the functions, performance, design constraints, and other attributes[93] of the system and its operational environments and external interfaces[94].

Before you can create the technical requirements, you must have business and stakeholder requirements. Whether you use the formal processes discussed in the preceding chapter (and suggested in Figure 4.7) or some less formal process, technical requirements engineering can only be performed with the possession of a concept definition. The concept definition encompasses the business or mission analysis results (business requirements), and the stakeholder needs and requirements definition results. Again, in some projects, you are completely unaware of how the concept definition was created (the processes used) and are simply handed a concept definition to begin the technical requirements engineering process. In such cases, it may be necessary to meet with those involved in the concept definition process to clarify some elements and assure accurate technical requirements development.

In most wireless IoT systems, the technology used is not just for technology's sake. The technology is there to support the business itself and, in most cases, has the sole purpose of making the business more effective through efficiency, risk avoidance, customer experience or reducing labor requirements, among others. As such, each technical requirement must link back to the concept definition. The concept definition (business, mission, and stakeholder requirements) is the foundation of the system definition (technical requirements). In the last chapter, we mentioned that each requirement will

---

[93] The attributes are the inherent properties or characteristics of the system that can be distinguished and defined. Some attributes may be inherent (not configured, but existing) and others may be assigned (configured). An inherent attribute would be any non-configurable property within the system. For example, if the system uses encryption with AES-256, but this bit level cannot be changed or configured, the bit level of the encryption is an attribute that is inherent. However, if it can be changed or configured, to AES-128 for example, it is an assigned attribute.
[94] IEEE 29148-2018 clause 3.1.33

have its own unique identifier, and this is commonly used to link (trace) technical requirements back to the concept definition.

Since each requirement has a tracing ID[95] as part of the definition process, each technical requirement should support and link to a specific business requirement or stakeholder requirement.

For example, we used "The panic button shall have service within the guest rooms and throughout the hotel." as a stakeholder requirement earlier. The business requirement driving the stakeholder requirement may be "A panic button IoT solution shall be deployed in compliance with local regulations." The stakeholder requirement can be linked to a technical requirement:

B.1.0   A panic button IoT solution shall be deployed in compliance with local regulations.
    S.1.1   The panic button shall have service within the guest rooms and other enclosed areas.
        T.1.1.1   The system shall provide a map or illustration of the hotel areas showing the coverage areas for the network.

In this tracing system, *B* indicates a business requirement, *S* indicates a stakeholder requirement, and *T* indicates a technical (or system) requirement. Let's assume Wi-Fi is the selected technology for this deployment. A secondary technical requirement could address the signal strength and requirements for coverage that provides accurate location services more specifically. This might read as follows:

The network in the coverage areas highlighted on the map in <T.1.1.1> must show at least 3 AP's visible to clients at a signal strength of -65 dB as measured by <selected panic button>.

The application shall determine the location of clients in the coverage areas highlighted on the map in <T.1.1.1> 99.9% of the time and this shall be accomplished the detection of client devices by at least 3 AP's utilizing tools available from the network equipment provider.

The more complete set of requirements, then, will look more like the following:

---

[95] We are using an example ID tracing system in this text based on hierarchical numbering, but any such system will allow for requirements tracing.

B.1.0   A panic button IoT solution shall be deployed in compliance with local regulations.

    S.1.1   The panic button <B.1.0> shall have service within the guest rooms and other enclosed areas.

        T.1.1.1   The system shall provide a map or illustration of the hotel areas showing the coverage areas for the network <S.1.1>.

            T.1.1.1.1   The network in the coverage areas highlighted on the map in <T.1.1.1> must show at least 3 APs visible to clients at a signal strength of -65 dB as measured by <selected panic button>.

        T.1.1.2   The application shall determine the location of clients in the coverage areas highlighted on the map in <T.1.1.1> 99.9% of the time and this shall be accomplished by the detection of client devices by at least 3 AP's utilizing tools available from the network equipment provider.

Note that in each of these requirements, enough specificity exists to allow for validation. If a requirement cannot be validated, you cannot determine success or failure. As we have covered before, requirements must be able to be validated. Any requirement that cannot be validated, cannot be considered a proper requirement.

While the preceding example shows the business, stakeholder, and technical requirements in one all-inclusive hierarchy, in your tracing system, you may have all the business requirements in one document, all the stakeholder requirements in another document, and all the technical requirements in yet another. However, the requirement IDs can still be used to link them.

The Process

The 29148-2018 defined process for system requirements engineering includes the following activities and tasks[96]:

1. Prepare for System Requirements Definition

---

[96] IEEE 29148-2018 clause 6.4.3

a. Define the functional boundary of the system in terms of the behavior and properties to be provided.
   b. Define the system requirements definition strategy.
   c. Identify and plan for the necessary enabling systems or services[97] needed to support system requirements definition.
   d. Obtain or acquire access to the enabling systems or services to be used.
2. Define System Requirements
   a. Define each function that the system is required to perform.
   b. Identify required states or modes of operation of the system.
   c. Define necessary implementation constraints.
   d. Identify system requirements that relate to risks, criticality of the system, or critical quality characteristics.
   e. Define system requirements and rationale.
      i. Data elements, data structures and formats, and database or data retention requirements
      ii. User interfaces and user documentation (information for users) and user training
      iii. Interfaces with other systems and services
      iv. Functions and non-functional characteristics, including critical quality characteristics such as security, availability, and performance
      v. Requirements attributes such as rationale, priority, traceability, test cases, information items, methods of verification, and evaluated risks
3. Analyze System Requirements
   a. Analyze the complete set of system requirements.
   b. Define critical performance measures that enable the assessment of technical achievement.
   c. Provide the analyzed requirements to applicable stakeholders for review.
   d. Resolve system requirements issues.
4. Manage System Requirements
   a. Obtain agreement on the system requirements.

---

[97] These *enabling systems or services* are not part of the IoT system per se, but are the systems used to develop the requirements. They may include collaboration systems, documentation systems, teleconference and video conference systems, and information repositories.

b. Maintain traceability of the system requirements.

4.3: Exploring Functional and Non-Functional Requirements

Functional requirements may stand on their own or work in conjunction with other functional requirements. That is, they may contain all the logic, hardware, and interfaces in a single requirement to accomplish a task. Alternatively, they may require interaction with other functional requirements, on some interface, to accomplish a task. These system capabilities, interfaces, operations, etc., are working toward the common goal of enabling the business requirements for a complete wireless IoT solution. The remainder of this chapter will explore various areas that should be considered when defining requirements beyond the wireless IoT network itself. These areas may not be areas of expertise for the wireless IoT designer, but the designer should be able to recommend solutions for exploration and understand how they fit into the requirements engineering process for the larger project.

System Capabilities

Our requirements need to be able to answer the question: What does the system need to be able to do from a functional standpoint? The goal of these requirements is to define what the major parts of the system itself will do. Here are some of the questions to ask and to answer as part of the requirements:

- Is resiliency required in the event of outages / failures?
- What happens in the event of a failure within different parts of the system?
- What network conditions should be expected?
    - Network latency
    - Bandwidth required between system components
- How will data flow across the network?
    - IoT Data protocols (e.g., MQTT, CoAP, HTTP, etc.)
    - Messages per day/hour/minute per device
    - Open or secure network
    - Mesh, star, tree, or another topology
- How will data be stored?
    - SQL databases

- NoSQL databases
- Textual data files (JSON, XML, etc.
- How many IoT sensors/nodes/end devices can be within a given area?
  - Distance between end devices or nodes
  - Distance from IoT gateways
- Is system automation a requirement?

The technical requirements around the system capabilities help ensure alignment among system components. The functional requirements allow you to get things done and the non-functional requirements allow you to get things done right. The demand for some non-functional requirements will themselves create functional requirements. Resiliency is a good example of this interdependency.

Resiliency[98] is an important consideration and, though it is a non-functional requirement, if a specific level of resiliency is required, functional requirements must be specified for it. The non-functional requirement will be the level of resiliency required. The functional requirement will be the implementation of a cluster solution, the implementation of redundant gateways, the implementation of redundant routers, and other components. We know that outages happen (scheduled or unscheduled), system maintenance must occur, data must go from system to system, and different locations (or areas) will have different connectivity requirements. The more resiliency required of the system, the more the costs of the project generally go up because resiliency usually requires redundancy and redundancy requires more hardware[99].

---

[98] Resiliency is a measure of the ability of a system to continue operations in the presence of failures within the system. Redundancy is implemented within components of a system to provide for system resiliency. For example, a subnet with two routers for connectivity to the rest of the network has redundancy in the routers and results in a more resilient subnet (the system) than a subnet with one router. Therefore, redundancy is typically defined at the component or element level (the components or elements of a system) and resiliency is measured at the system level.

[99] In a well-engineered system, the redundancy costs are recuperated through the increased dependability of the system. For example, if no redundancy is implemented and a critical device fails that disables the entire IoT network for six hours, the organization could lose many thousands (or millions) of dollars because of the incident. Spending a few thousand or a few hundred thousand dollars, depending on the size of the IoT solution, can actually save the organization many times more in the end. Therefore, when planning for resiliency, it should be a factor of downtime cost and not based on the desire to simply have it.

Defining what happens in the event of a failure ensures that critical data is not lost. Some data may only provide value in real time, while other data sets are required for trending. If the data is only valuable in real time, storing this data on the sensors to send it later, in the event of any systems being unavailable, may not make sense. The reverse is also true, if the data is important from a historical perspective, you will need to store this information and forward it later. This assumes that no security protocols are being breached by queuing the data.

Several of the decisions here will directly impact the decision for network connectivity in the IoT solution. For example, if your network will potentially consist of hundreds of nodes with hundreds of messages each per hour, SigFox or LoRaWAN probably are not going to be the best connectivity option for you, depending on how LoRaWAN is implemented. Latency requirements will also drive this. You can determine the amount of bandwidth required by understanding the message flow. How many messages are sent and what is the size of each message? Then, factor in the number of end devices using this message structure and you will understand the bandwidth required[100]. A non-functional requirement related to bandwidth might be:

The wireless IoT network shall provide for a minimum of 20 kbps per end device and provide transmission of a maximum of one megabyte per hour.

What are the security requirements for connecting to the network? Is an open wireless connection all that is needed or is something more secure required? Sometimes moving away from commonly used protocols and frequencies can decrease the likelihood of a security breach due to a lack of tools available to decrypt these connections being generally available. However, for critical systems, such "security through obscurity" means should not be relied upon alone. Instead, authentication to the network and encryption of transmissions should be used.

When designing the IoT system from a vendor perspective you need to think about the common situations in which the customers will find themselves. This means thinking about how many nodes could be on a single network, what services those nodes would

---

[100] One of the advantages of wireless IoT design is that most end devices serve a single purpose. You are not as frequently concerned about devices that have varied applications and use one amount of bandwidth one day and another amount on another day. Instead, the devices are typically consistent in what they send. Some devices are random as to when they transmit but are still very consistent as to what they transmit.

require, and whether some areas have more nodes than others. For technologies that are spread out over great distances, LoRaWAN and Sigfox may make the most sense with limits of 10,000-15,000 meters, but if you are doing wayfinding within a stadium or arena something with less range like BLE could make more sense.

Interfaces

Understanding the interface requirements is critical for a data driven IoT environment. The interfaces within the IoT system and, if applicable, interfaces to other systems are important. If the IoT system itself consists of more than one component and you desire communications between these components, interfaces need to be defined. Broadly, IoT interfaces fall into three categories, connectivity (network connection), the primary user interface to the IoT system (human-computer interface (HCI)), and systems interfaces. Depending on your role when designing the IoT system, you may not get involved with the inner workings of the IoT system interfaces, but you should specify them in the requirements and in your design so that the proper professionals can ensure they are implemented.

With most things within the system, a flow diagram can help logically show how data will flow through the system. Figure 4.8 illustrates the concept while identifying interfaces between systems.

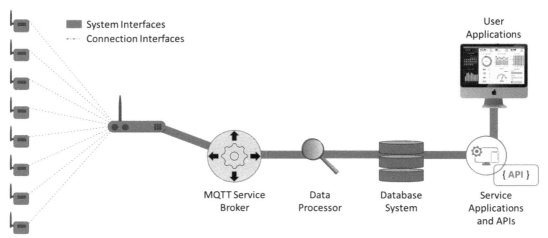

Figure 4.8: System and Connection Interfaces

Connectivity

The way the IoT sensors will connect to the network needs to be determined and defined. There are many options to choose from for IoT devices and topologies that may be utilized (the next chapter covers topologies and architectures to provision you for the proper selection during design or understanding when they are constrained upon you). The goal for this chapter is to create technical requirements for connectivity and not to review the details of why one technology would be selected over another.

To define the technical requirements for connectivity for the IoT sensors, it helps to start by asking a few questions to determine if mobility is a concern:

- Are the IoT sensors going to move?
- If so, how frequently will the IoT sensors move?
- If so, how far will the IoT sensors move?

These will help define the broad category of mobility connectivity options for the requirements. If mobility is required for IoT devices in a relatively small area (10,000 square meters or so), most any wireless IoT protocol can be implemented with sufficient gateways to achieve the goal. If a much larger area is required, LPWAN protocols may be needed such as LTE/4G/5G, LoRaWAN, or Sigfox.

Will the sensor be part of a PAN, LAN, MAN, or WAN? Utilizing a classification system can help narrow down the options to the best connectivity option for an IoT project. After this, building the requirements can focus on the connectivity protocols that match with the appropriate connectivity classification. High-level classifications might include long-range and short-range. Within short-range, classifications might include industrial or commercial versus consumer. Such classifications of protocols, defined within an organization, can assist in more easily selecting the proper protocol for an application.

Capacity is also an important piece of the planning and selection process. Here are some common constraints or system attributes that need to be considered:

- Number of devices
- Messages per day per device
- Message Size
- Distance
- Data rates

If the requirements are gathered around these connectivity items, it will be easier to select the appropriate connectivity option when the project reaches the Design phase for requirement fulfillment.

User Interfaces

The user interface to the IoT system itself can take on many forms. Common system interfaces include IoT Smart Phone applications, HTTP(S) interfaces to a website, and purpose-built software and hardware. For example, an IoT thermostat doesn't require a computer application, it can simply provide a touch screen interface on the device. However, it may also make available a mobile phone app or web-based interface for management on a large scale for enterprise and industrial deployments. The goal is to create technical requirements for the user interface that align with the business requirements. Ideally these requirements should be straight-forward and not allow for misinterpretation. Figure 4.9 is an example of a device with a non-user-friendly interface; however it is not intended for "daily" use but rather to connect other devices to a wireless network. It acts as a gateway for wire-only devices of industrial use, such as thermostats and other BACnet and Modbus devices. Figure 4.10 is an example of an interface more tuned for user needs, but it offers a complex industrial-grade capability on the backend.

Figure 4.9: Babel Buster Series BAS-7050-RT

Some IoT systems will require a user interface on the IoT sensors themselves, like the example thermostat in Figure 4.10. In other cases, the sensors and other devices are just data collectors or control actuators with no direct user interface, except for some LEDs like the example in Figure 4.9. These devices will send the data upstream to the IoT data collector for information to be processed accordingly. Once the data is processed it can then be displayed on any type of user interface the system may provide or that you may create based on the IoT data.

Figure 4.10: tekmar 564 Invita WiFi Thermostat

When defining the user interface requirements, here are a few questions that are helpful to ask:

- Who is the target user for each interface?
- How will they access the interface?
- Are they accessing it using a device they already have?
- How simple/complex does the interface need to be to be effective? For example, are the users engineers or business professionals?
- Is the information being provided valuable to the user?

Answer these questions and then you can start drafting the actual requirements for the user interface. The user interface in some IoT systems may be more complex than our thermostat example or even require constant monitoring. In the case of the hotel panic button the user interface for the housekeeping is a button; however, the security team may be monitoring a video system and may receive a pop-up alert. This is also a user interface and could require an acknowledgement of some type. In any case, a notification will occur in some user interface, be it a text message on a cell phone or an alarm on an internal communication system.

System Interfaces

System interfaces cover both internal and external interfaces that are part of the IoT system. Internal interfaces communicate as part of a closed system and provide data or services from one system to another. External interfaces communicate with other systems and provide data or services.

Depending on your role in designing the IoT system, you may not have visibility into the external interfaces between systems. In this case, it can be treated like a "black box" and the requirements you create would focus on the data that you can get out of the system and that can be transformed, if required, into data that you can use. Stated differently, your requirements simply state that the system shall provide x data or x service to the system of interest. The only necessary information is the network connection method used between the systems, the format of the data received, and the method for requesting the data. How the external system gets, generates, or forwards the data is not important to you. Once received, you may have to convert the received data into a format compatible with the system of interest. These conversions are generally handled with scripts, internal application code, or an ETL (extract, transform, and load) solution if large or complex conversions must be performed.

Common interfaces for external system functions or services include technologies like REST APIs, MQTT servers, service APIs, etc. Examples of common IoT Data protocols, including MQTT, AMQP, CoAP, XMPP, and DDS may be used. CWIIP covers the protocols in more detail and how to integrate with them. For the purposes of the CWIDP certification you need to understand how these protocols function and when you might use one over another option to meet your requirements. They are discussed more in later chapters.

Here are some questions around external interfaces that need to be answered as part of the requirements engineering process.

- What data will be made available?
- Who or what will need to consume this data?
- How does the system maintain confidentiality of the data?
- Is the data in the correct format or does it need to be converted?
- Does a method already exist to provide the required data?

Answering these questions is a solid first step in creating requirements for external interfaces. If a method to provide the data doesn't exist, the manager of the system may have to be involved so that a method can be developed. Exporting data to a folder and then processing new data with a script has worked for years in technology and still works today; however, if the system can integrate with common IoT Data protocols, the process is more stable and efficient.

System Operations

Ongoing maintenance and management are generally required for technology solutions. The scope of ongoing management encompasses the IoT sensors and the systems required as part of the overall IoT solution. There should be discussion with the business on how much ongoing maintenance is acceptable and who will perform that maintenance.

In general terms, the less maintenance required for the IoT sensors, the more expensive the solution will be. Many IoT sensors are battery powered and these batteries will need to be replaced on some ongoing basis. Long-life batteries are more expensive that others. Therefore, less maintenance equals greater expense. Additionally, industrial grade enclosures for IoT devices raise the price and reduce the maintenance. You get the picture.

When considering the batteries, it is important to define what is an acceptable interval for battery changes and plan the technology accordingly. For example, having a Wi-Fi radio powered 24x7 may not be the best option for a battery powered sensor that won't be charged on a regular basis. Think about how long a mobile phone holds a charge when you are using it? A purpose-built sensor that is designed to stay in a location without power can only last so long. Though, with the right hardware, protocols, and batteries, ten years is not unreasonable.

Beyond the Exam: BLE, where are you?

In one deployment we were using BLE beacons to assist with wayfinding within an arena. The individual who knew the exact locations of the sensors left the company. No one realized we didn't have documentation about the location of the sensors until the batteries on the sensors started dying and the application stopped being useful. This deployment took months to plan and due to lack of documentation, we had to deploy a

> team to go on a scavenger hunt for about a week to find them all, document them, and replace the batteries. Finding the roughly 250 sensors to replace the batteries would have been much easier with proper documentation located in a central repository. Interestingly, we could not use technology to locate the locators because the simplest of technology had not been used: documentation.
>
> *-Darrell*

Other environmental conditions require different considerations. We have talked about the NEMA ratings earlier; however, in these harsh environments it is important to ensure the sensors themselves maintain protection from the elements. Weather-proof seals will rot or crack over time if no maintenance is performed. For any system that is critical to the ongoing state of the business, catching a potential problem before it becomes a problem is good for business.

On the systems side, maintenance is different from the sensors themselves. Just like the sensors, the condition of the hardware is important; however, if you are utilizing a cloud or managed server environment this may be less of a concern. If the equipment is onsite routine checks should be done to ensure the electronic components are still operating within the manufacturer recommended guidelines for temperature, humidity, etc.

In addition to the physical side of ongoing maintenance there are several logical things that need to happen. Here are a few things to consider related to operations:

- Operating System
    - Backup / Restore
    - Log Rotation
    - Troubleshooting
    - Security patches
    - Upgrades
- Software upgrades
    - Alerts & Notifications
    - Backup / Restore
    - Bug fixes
    - New features
    - Troubleshooting

We are not saying this list is inclusive of everything from a maintenance standpoint; however, it is a good non-industry specific starter set. Being able to configure which alerts and notifications are provided is extremely helpful, as different groups may handle different pieces of the system. Replacing a hard drive failure on a RAID array is probably not completed by the same person who is changing the batteries on the sensors that have gone offline in a large, distributed deployment.

Data and Event Collection and Control

In IoT environments, as has been stated several times already in this book, data is the primary focus. Whether it is data sent to an actuator so that actions can occur, or data sent from a sensor so that machinery and other devices can be monitored and controlled, it comes down to the data. The purpose of the entire solution is to get the right data to the right place in the right format at the right time.

ETL Processes (Data Collection Parameters)
Extract, transform, and load (ETL) processes have been used in database systems for decades, quite literally going back to the earliest days of database systems, though it wasn't always called ETL. The names are self-explanatory and provide services to pull the desired data from one location, massage it into the format or construct desired and possibly create additional calculated values, and then import it into the target location.

Data transformation is a significant part of data requirements and is a part of system operations. Particularly in IoT systems, you may have data entering the system from the IoT devices using many different formats. For example, I once worked with an IoT solution that used temperature sensors. They had deployed three very different sensors. All three reported temperatures using a different data representation model. While not the exact model, consider the following as an example:

- *System One:* Temperature reported as a decimal value between 1 and 254 (one byte, 0 is used to indicate no reading) based on the lower range of -30 degrees Fahrenheit and the upper range of greater than 200 degrees Fahrenheit.
- *System Two:* Temperature reported in Celsius using one byte with decimal values equal to the temperature.
- *System Three:* Temperature reported only as safe or unsafe based on the system configuration of lower and upper thresholds.

As you can see, this data would be coming in using very different formats. You may require that the data be standardized before storage in the database system. For example, the first two data values could simply be reformatted to match each other in a standard destination format, Celsius for example. Therefore, system two would already be in the right format and the data could simply be transferred. However, system one would require data transformation, which would covert the Fahrenheit values to their equivalent Celsius values and then store them in the database system. To accommodate system three, you may allow for a specific temperature value or a range of temperature values to be stored in the system. When the sensors report "safe", you could store the range in Celsius that is known to be the configured range. When the sensors report "unsafe", you could store a value in the database system that triggers an alert. For example, if you know you are unlikely to see the value 1000 in Celsius, you could use that value for the trigger value.

In the end, you have a system that aggregates the data and reformats it in a consistent way that will work with the current IoT sensors and, when designed well, with future IoT sensors as well.

The process of ETL is what we have defined here. Special tools exist to assist with this, and custom code can be developed to accomplish the task as well. For example, many IoT programmers use Python today to achieve ETL operations without expensive database management tools[101]. Big data platforms, such as Hadoop, are very focused on ETL because data ingress and egress are the lifeblood of such systems. ETL is effectively about defining, implementing, and managing data flows. Getting data into the proper storage (ingress) and out of the storage for applications (egress) is the fundamental goal.

No Data Storage
To be as accurate as possible, we must address the concept of data in relation to events as well. Particularly with event systems – those that monitor something and report only when an event occurs – it is not uncommon to have no permanent data storage. The IoT sensors report events that occur, and the system takes immediate action. No user intervention is required. In such cases, many still choose to implement storage of the

---

[101] To avoid misrepresenting the case, some tools are expensive, and some tools are free. Yet, others come with database systems. For example, both SQL Server and Oracle provide an ETL tool for their database systems. Several open-source tools are also available to assist with ETL and ETL-like operations including Talend Open Studio, Singer (a command line ETL toolset), AirByte Community, and Apache Camel.

events for historical records, but I've seen a significant number of systems that stored no data at all. With that said, and it is an important consideration (always ask, "Do we need to store this data value?"), let's discuss data storage requirements.

Data Storage

When it comes to data storage for IoT, the decision can be critical. In some environments, sensors are reporting readings every few seconds. If one thousand sensors report a reading every five seconds (some do it more frequently in some implementations), that's 12,000 reading per second, or 720,000 readings per minute, or 43.2 million readings per hour, or 1.0368 billion readings per day. One facility, one thousand sensors, and one billion readings per day. There are, at least, four common ways to deal with this volume of traffic, depending on the requirements of the system[102]:

- Only store readings that indicate a problem.
- Only store the average of all readings every <period> and every reading that indicates a problem.
- Store everything.
- Store nothing.

The first option removes significant amounts of data that could be useful for business analysis – you may simply not know it yet. The second option provides at least some data for statistical analysis. The third option will require an excellent data storage solution. The fourth option, while optimal for performance and minimizing data storage, provides no long-term analysis options. Whatever decision you make during requirements engineering, you should be aware of the data storage options so that you can ensure the requirements created can be fulfilled.

---

[102] Certainly, there are more ways, but they are not as common. One example is to analyze trends in a solution and store only changed data. For example, an IoT sensor may be sending reports every five seconds that, on average, contain the exact same value ten or more times in a row. If that's the case, we could store the value with a time stamp and then only store another value when it has changed and with a new timestamp. Given that we know the reporting interval of five seconds, we can determine the other values that would have been transmitted between the timestamps. This is not really imputation or even interpolation, as we do know what the values were, we just didn't store them. If the values had changed, a record would have been created. I've seen several other methods for dealing with the IoT Really Big Data (my term) problem, and I'm sure you will encounter and create more methods.

Before we explore data storage solutions, consider the three dimensions of data analytics and the factor they will play in your requirements. The first dimension is *the past*. What happened? Why did it happen? When did it happen? Where did it happen? How did it happen? This dimension is about the analysis of historical data. IoT devices cannot provide this information, they already provided it. What I mean by this statement is that it is up to the IoT solution as a whole to provide historic data. All the IoT devices can do is tell your system what they just sensed[103]. Providing historical data analysis requires storage and, if you want detailed historical analysis, it requires lots of storage in the world of IoT.

The second dimension is *the present*. This dimension is about real-time analysis. What's happening now? Why is it happening? Where is it happening? How is it happening? This is where streaming data analysis comes into play. We often create live dashboards that report the values, typically in a graphical manner, as they are coming in. Think of this dimension like a pressure gauge on a tank. You can walk up to the gauge and view the pressure as it is measured at this immediate moment. An analog-only gauge will not show you historic information, but it shows you real-time information.

The third dimension is *the future*. This dimension is about predictive analysis. What will happen? Why will it happen? Where will it happen? When will it happen? How will it happen? Your IoT devices cannot provide this to you at all (well, in most cases, some devices can do on-board processing and send prediction values to the network, but more often, this is done within the larger IoT solution). Predictive analysis depends on stored data from the past, the discovery of trends, and the determination of where the trends are leading.

Sometimes, you require only one of these dimensions. At other times, you require all three. It is essential to elicit the needs in this area to ensure that you recommend a proper data processing and storage solution. The organization will see you as the IoT expert and you should know enough about the data and event collection parameter options to elicit the appropriate requirements and recommend possible solutions in your design. You will not be an expert in Hadoop and Splunk and MySQL and SQL Server, etc., but you will know what use cases they serve best so that you can recommend the right technology. If

---

[103] Some devices may retain a few historic values that can be queried, if required. However, the normal processing within IoT solutions is for the end devices to send the data through the network at some interval and it is then stored for analytics.

you want to learn more details about these data storage solutions than that presented in this book, which is sufficient for design work, consider the CWIIP Study and Reference Guide or select a good book on Hadoop or another big data solution.

The following are common modern data storage and analysis solutions useful for IoT data[104]:

- **Hadoop:** An Apache project that provides a framework for distributed processing of large data sets (Big Data) using clustering. It can run on a single server or scale to thousands of computers with local compute capabilities and data storage facilities. If uses the Hadoop Distributed File System (HDFS) to provide access to the data that exists on multiple nodes in the cluster. It is also well-known for its use of MapReduce, which provides the parallel processing within the system. Hadoop is useful for large-scale IoT projects with large data volumes, high data generation rates, and variable data types or structures. Hadoop is not a database; it is a framework that supports multiple database types including SQL or relational databases and NoSQL and unstructured databases.
- **Apache Spark:** Another Apache project that provides a framework for distributed processing with intentional design for stream processing (real-time processing) as well. Generally considered an alternative for Hadoop and optimal for real-time analysis solutions. One reason Spark is more efficient for real-time processing is that it uses in-memory data more than Hadoop and can perform from three to one hundred times faster than Hadoop on many processes. In both Hadoop and Spart, the data is sourced outside the system and may be stored in the system for processing purposes. In many cases, Hadoop and Spark are used together. Hadoop is often used for storage with HDFS and Spark is used for compute processes against the Hadoop data sets.
- **MongoDB:** MongoDB is a document database, which means that it stores document objects instead of data records (like a traditional relational database, such as SQL Server). The document objects should not be thought of in the common user application sense, like a Word document or another application document. Instead, the document object is an unstructured data object. A

---

[104] This list is not exhaustive but introduces many of the commonly used data store and analysis solutions.

structure may exist within the document object, but the MongoDB database need not be aware of this structure. Stated differently, structure, if used, is imposed by the application and not the database management engine. An example document is shown in the image below:

```
{
  _id:"4098098789",
  sensorID:"01003",
  sensorType1: "Temperature",
  sensorType2: "Humidity",
  sensorFormat: "Celsius",
  sensorLocale: "First Floor",
  sensorReadings: {
    temperature: "22",
    humidity: "63"
  }
}
```

Such databases are useful for unstructured or semi-structured data and data writes are typically very fast as well. MongoDB and other document databases[105] and NoSQL databases are common in IoT solutions.

- **DocumentDB:** The AWS version of a document database that is compatible with MongoDB through implementation of the MongoDB APIs so that applications written for MongoDB work with DocumentDB in the AWS cloud.
- **Firestore:** The Google cloud document database solution.
- **Cosmos DB:** The Azure cloud document database solution.
- **Cassandra:** A NoSQL database that uses tables, but it uses them differently than a traditional relational database in that Cassandra is column-oriented rather than record-oriented. First, the tables can be deployed without an architecture called a schema that defines the whole structure of the database. Second, Cassandra does

---

[105] A document database is one type of database in the category of NoSQL databases. A NoSQL database is one that does not adhere to the traditional relational database model that goes back to the 1970s. Instead, the data is stored in some other format, such as document stores, key-value stores, graph networks, and column-oriented databases. A common characteristic of these database types is the lack of a fixed schema. Instead, with the document database as an example, each document can have a different number of key-value pairs within the JSON object. For example, one may have sensorType1 and sensorType2, while another has only sensorType1 and yet another has an additional sensorType3. The structure comes from the applications rather than the database engine. Therefore, the application must check for sensor types, in this example, until no more entries exist.

not support relationship and therefore all data needed to service a query must be in a single table (in most cases and depending on how you define *query*). The design of the database is called a query-based design and it is intentionally a denormalized data store. Therefore, the entire data store will be much larger than a relational database but reads will be very fast. Therefore, it is an excellent solution for read-based analysis of data and is often used as a final destination for data in IoT solutions.

- **DynamoDB:** An AWS NoSQL database that stores key-value pairs in tables (though they are conceptually documents) and, like Cassandra, it is not relational in nature. DynamoDB is often considered a document-oriented store with key-value pairs, while Cassandra is considered a column-oriented store.

The following are common traditional relational database storage solutions[106] (Figure 4.11 shows the ranking of the top four relational database management systems):

- Oracle
- MySQL
- SQL Server
- PostgreSQL

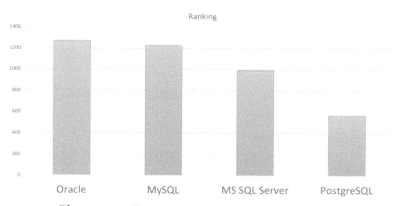

Figure 4.11: Top Four RDBMS (2021, Statista[107])

---

[106] This list is not exhaustive but introduces many of the commonly used relational database and other storage methods.

The following are methods of storage that use no real database:

- **Log File Storage:** Many IoT systems generate log files. The logs may include data from sensors or simply operational logs. Such data may be useful for analyzing the health of the system in general. It can be pulled into other database systems for analysis and presentation. Many solutions, discussed later, are designed specifically for processing log file data. Log files may be stored in text formats or binary formats, depending on the system.
- **Text File Storage:** The simplest (and one of the oldest) storage methods is text file storage. Test formats include XML, JSON, HTML, CSV (Comma Separated Values), TSV (Tab Separated Values), and others. They are among the simplest formats to ingest into an application, but they are far less efficient for large data sets. However, in small-scale IoT deployments, they may have their use.

Data and event management is about more than just retrieving and storing and analyzing the data. It also includes data presentation based on analysis. Custom applications can perform the task, but solutions also exist that can be modified or configured to perform the role. Many solutions originated in system log analysis and security information and event management (SIEM) and have been heavily used for IoT monitoring. These include:

- **Splunk:** A commercial platform for data analysis and presentation. It includes the ability to search, index and process data and display results in dashboards. It also supports alerting. Commonly used in the information security industry and DevOps, it has been used some in IoT analytics as well.
- **Grafana (Prometheus):** An open-source alternative to Splunk, Grafana is useful for and frequently used in IoT analytics. Grafana is the visualization tool and Prometheus is most commonly used with it to collect and process the data. However, it is in no way limited to Premetheus. In fact, AWS offers compatibility between its IoT SiteWise solution and Grafana. They even offer the AWS Managed Service for Grafana (AMG) where you can host both the harvesting

---

[107] It is interesting to note that the most popular databases do not equal the most wanted database skills. The market is full of people who know Oracle, MySQL, and SQL Server. Instead, the most wanted skills, according to Statista from 2021, are PostgreSQL, MongoDB, Redis, Elasticsearch, MySQL, Firebase, SQLite, Cassandra, DynamoDB, MS SQL Server, MariaDB, Oracle, CouchBase, and IBM DB2 in that order.

and visualization solution within AWS. Figure 4.12 shows an example implementation with Grafana.

- **Knime:** The Konstanz Information Miner (KNIME) is another open-source solution that is considered "end-to-end" as it supports everything from the ingestion of data to the end visualizations. Connectors are available to connect to many data sources including some IoT-specific solutions, such as Cumulocity IoT, which is a primarily cloud-based solution for centralizing all IoT data without the need for complex programming.

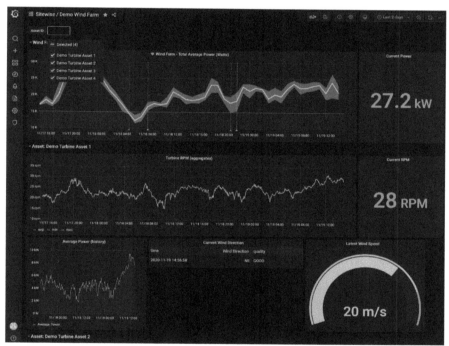

Figure 4.12: Grafana Showing IoT Reported Information[108]

- **RStudio:** While RStudio itself is primarily an integrated development environment (IDE) for the R language, it can be coupled with the Shiny Server open-source solution to publish Shiny web applications, which can provide visualization of IoT and other data. R is a language used heavily in data science

---

[108] For more information on this example, see docs.aws.amazon.com/iot-sitewise/latest/userguide/grafana-integration.html

and it is making significant inroads into the IoT space due to its usefulness in analyzing the massive data generated by IoT devices. RStudio provides the development environment and Shiny provides the publication service for reports, graphs, and other web applications.

- **Kibana:** Another open-source solution that provides visualization of data. It is specifically designed to work with Elasticsearch (produced by the same group) and works well with the Elastic Stack, discussed next.
- **ELK(B):** Elasticsearch, Logstash, and Kibana (ELK) (often used with Beats as well), called the Elastic Stack, is an open-source suite of tools commonly used together to find needed data in logs or other data sources and present the results visually in a web-based platform[109]. While this solution may seem more complex than others, due to its use of multiple components, the setup and configuration is actually very well planned and involves only installing and configuring each product and then connecting data sources and generating visualizations based on analysis. To make it even easier, Docker images are available to quickly deploy the entire stack. Of course, it is a powerful solution and, therefore, many features exist, which will require some initial learning.

Clearly, the options for data storage, analysis, and visualization are extensive and those reviewed here represent commonly used solutions, but many others are available as well. While many of the solutions referenced were not designed specifically for IoT data analysis, they work exceptionally well with it and the fact that many cloud service providers seek to integrate their IoT platforms with them shows this reality.

Programming Languages

Programming languages are likely to be used in an IoT system. They may be simple drag-and-drop interfaces that build scripts behind the scenes (such as those used in some ETL solutions) or you may start with a blank screen and begin writing code. Either way, defining requirements related to programming languages is important to improve supportability and longevity of the integration system. These programming languages include compiled languages as well as scripted languages.

Selecting or requiring specific programming languages is often driven by these factors:

---

[109] Logstash and Beats are used to get the data into Elasticsearch and then Kibana provides the access to, analysis of, and visualization of that data.

- What languages are already used in the organization with individuals who understand them?
- What languages are required for the alternate solutions you have identified up to this point?
- What languages are best suited to accomplish your goals?
- What languages can be used in compliance with your security requirements?
- What languages are supported/required by the APIs provided within your IoT systems (those being deployed) or other systems that already exist are will be implemented in the IoT solution?

As you can see, it's not as simple as choosing the language that you like the best. If that were the case, everyone would use Python for non-compiled scripts and C++ for compiled applications (no author bias here). The reality is far more complex than this. You are often constrained by what is supported within the systems. The good news is that more and more IoT systems are moving to RESTful APIs, which means that you can use any language that can call a RESTful API – which is most languages.

From a requirements engineering perspective, it is simply important to identify required languages that should be supported by the IoT solution. That is, the languages for which the IoT solution must provide interfaces.

Non-Functional Requirements of Importance

Again, non-functional requirements are requirements that enhance functionality but to not directly provide functionality themselves. Stated another way, they enhance your ability to do things, without actually doing those things. This section will address several important non-function requirements. These include requirements such as the following:

- Usability
- Security
- Performance
- Reliability
- Scalability
- Coexistence

If you are designing industry specific IoT systems, you may find additional areas that are specific to that industry or as part of your brand identity. As an integrator or implementor of IoT systems, the ones mentioned above should cover most use cases. Now, let's dive into each of these items in a bit more depth.

Usability

Defining how to use the system is probably one of the most important system quality requirements. When you think about usability it is more than just how the primary user will connect to the system. For the IoT system to be successful, requirements around how users will use the system are important. When you think about successful products, they tend to be easy to use.

Your IoT system doesn't have to be as easy to use as a simple system; however, defining and understanding how it will be used and discussing this with stakeholders ahead of time can reduce the possibility of failure. For example, if the device is meant to be carried on a person, having something that is large and heavy probably won't be very usable for most people. Likewise, if the system is battery operated these batteries need a long lifespan or the maintenance cost may be unbearable, in this example the usability requirement of long battery life could drive specific IoT backhaul choices.

When creating requirements around usability make sure you cover each type of user and how they might use it. Creating a list of users can help in ensuring no one is missed.

Security

The CIA security model, confidentiality, integrity, availability, is often the beginning of discussions on security. These three pillars are vital to your security in IoT as well. When determining data requirements for your IoT solution, always ask the following questions related to the CIA security model:

- How will we provide confidentiality? This is keeping the information away from unauthorized users and allowing visibility only to those authorized.
- How will we provide integrity? This is ensuring that the data received is the data sent (in the network communications context) and ensuring that the data is stored without corruption (in the data storage context). It ensures that the data has not been modified without authorization.
- How will we provide availability? This is ensuring access to the data is there when it is required, by whom it is required, where it is required, and how (though this may be more of an accessibility issue depending on definitions) it is required.

Most things in security relate back to these three pillars. Each of these have several more questions that need to be answered to create your requirements, but the above gives you a starting point.

When you think about confidentiality, for example, this includes ensuring a user only sees the data views that the user's account is authorized for. The more sensitive the data being collected is, the more security mechanisms should be put into place. For example, if you are doing anything with credit or debit card processing, you will have to ensure the system follows the rules laid out in PCI DSS. Different types of data have different levels of sensitivity, and the business should communicate which requirements need to be followed around the data sets to the individual creating the technical requirements. Your job is to translate these policies into obtainable architectures and to design or implement a system that meets these requirements.

Depending on the level of confidentiality required, you may elect to keep the IoT wholly contained onsite on a closed network, not electing to utilize outside services such as cloud services. Preventing the data from leaving the local network is one method of securing the data from hackers; however, with all things in security it is a best practice to use a layered defense approach. This approach indicates that the transmission should be secured even if they are only on the local network, only required data should be transmitted, devices should be physically secured, etc.

Ensure data integrity is included as part of the security requirements as well. In a multiple component system, how do we know that the data we send from the sensors is received by the target only and not intercepted along the way? Will acknowledgements be required? Will these be part of the underlying IoT protocol or are upper layer acknowledgements required? Will integrity checks be included in data storage and analysis processes? How will we control or limit the access of human users to modify machine generated data?

The requirements around availability of the data leads into reliability and scalability, which we will cover in more depth later in this chapter. For the purposes of this section, we focus on access to the data. It is part of requirements engineering to determine who needs access to what dataset and how will it be delivered.

Performance
When defining the system performance requirements and parameters it is good to start with the end-to-end expectations and then work backwards for the interactions within

the system. The performance expectations set here will directly impact the selection criteria for scalability and influence the costs of creating resiliency as part of the overall system. For example, if your best connectivity option to the Internet is 200 ms, having a system that expects 20 ms or better latency would likely prevent direct use of the cloud and a local option would be better.

Common considerations for performance are:

- Response times for each interaction
- Uptime
- System load
- Memory usage

Once the end-to-end performance expectations are set, each area of the system will need to have its own performance service levels associated with them. These can and should be defined in the requirements document.

Reliability

As a requirement, reliability is often overlooked or just assumed. But understanding the real requirements around reliability and balancing those against the project budget is an important exercise. Do the water sensors in the field that gauge the soil moisture really need five nines of reliability? What would the cost be to ensure such a reliable network? Would 99.999% uptime create a cost to deployment that is unbearable? On the flip side, having a system that isn't available when it supports safety isn't an option.

Reliability starts with the IoT sensors (the devices themselves and the reliability of the components in them) and extends to the network, system, and interfaces. Ideally each area should have its own individual performance and reliability metrics. For the IoT sensors, this means defining how they will be powered, will they be powered from the grid, Power over Ethernet, or be battery operated? Each option has a different implied reliability rating for the device having power.

Similarly, the IoT sensors need to have the appropriate climate protection if they are targeted for areas that are exposed to the weather. Areas near the ocean tend to have a saltwater spray that can corrode metals and destroy electronic components. The IP Code can be helpful in determining the requirements for your project and can be found here: https://en.wikipedia.org/wiki/IP_Code

From a network standpoint the reliability requirements may drive a decision to eliminate possible connectivity types. If you need a highly reliable connection, the congested 2.4 GHz space may not be a wise choice unless the protocol has resilience and can operate around other 2.4 GHz protocols. It may be better, in some cases, to look at other frequency bands as an alternative and protocols that operate in them (LoRaWAN, Sigfox, Z-Wave). Having sensors at the edge of the connectivity range may also be problematic with changing environments that could impact the signal strength enough to hinder connectivity periodically.

Having the target systems highly reliable helps ensure no loss of data; however, if the data is extremely important it may be wise to queue the data or confirm delivery of the stream itself.

Another perspective on reliability is to define it specific to the times in which it is required. For example, do not use the following requirement statement:

> *The IoT devices shall have connectivity to the network and uninterrupted data transfer capabilities between the end devices and the target data processing application 99.99% of the time.*

Instead, define the time windows in which this requirement must be met:

> *The IoT devices shall have connectivity to the network and uninterrupted data transfer capabilities between the end devices and the target data processing application 99.99% of the time during manufacturing hours.*

Then, elsewhere, define manufacturing hours. This method allows for repeat use of the phrase *manufacturing hours* without requiring a complete sentence- or paragraph-level description each time. For example, manufacturing hours may be defined as:

> **manufacturing hours:** *the window of time each day, which may vary according to the time of year, in which manufacturing processes are active within the organization.*

Why is it important do define the reliability requirement time window? To ensure proper interpretation of the requirement. For example, the first requirement did not constrain the statement to the hours of manufacturing activity. Therefore, it would allow for .01% downtime each year. This would allow for about 88 hours of downtime per year and the operations team may assume that these 88 hours can be taken from the active manufacturing time. The second requirement could not be interpreted in this way.

Assuming an average of 14 hours of manufacturing time per day at five days per week, the second requirement would only allow for about 35 hours of downtime per year, specifically during manufacturing hours. This is just one example of why it is so important to ensure requirement statements are written with clarity.

The suggestion to have a definitions section in the requirements document is also beneficial. It should be targeted at both clarifying any terms or phrases that could be interpreted other than intended and at reducing the overall size of the document. For example, the following three requirements can all use the phrase *manufacturing hours* without further defining the phrase, if it is defined in the definitions section:

> *The IoT devices shall have connectivity to the network and uninterrupted data transfer capabilities between the end devices and the target data processing application 99.99% of the time during* **manufacturing hours***.*
>
> *System maintenance tasks shall be scheduled outside of* **manufacturing hours** *except those maintenance tasks required for critical security and operations quality that must be performed immediately.*
>
> *The data merge process shall be performed once every 24 hours outside of* **manufacturing hours***.*

If the definitions section is not implemented, the last requirement (and the other three) would require replacing *manufacturing hours* in the following way, which would be far less readable and make for a much larger document:

> *The data merge process shall be performed once every 24 hours outside of* **the window of time each day, which may vary according to the time of year, in which manufacturing processes are active within the organization***.*

Finally, in the example requirement statements above, the terms and phrases *connectivity, end devices, critical security,* and *operations quality* should be defined and reused throughout the document for the same concept.

Scalability

If you are designing a solution to be sold as a product, scalability is extremely important. It is also important when you either do not know the future scale or do not desire to pay for the future scale today. Cloud solutions are often beneficial in the area of scalability, but scalability can be achieved in on-premises solutions as well – the hardware is just much more expensive.

Scalability is achieved when a system can grow to meet future demands. Scalability is similar to elasticity. Most people think of scalability as the ability to grow a system[110]; however, the technical definition of scalability is simply the capacity to be changed in scale or size[111]. Elasticity, more clearly, allows for growing or shrinking depending on what is required. You could define elasticity is dynamic scalability. For example, in cloud solutions, you can implement dynamic provisioning that ensures a resource is there when you need it, but it is not there (and therefore incurring costs) when you do not need it. An example might be a server instance that is required to run a process once each week. The instance can be provisioned, and the process can be executed (for however many minutes or hours required) and then terminated. This dynamic provisioning is an excellent benefit of cloud architectures, whether private clouds or public clouds.

Three basic architectures are available for hosting your such systems:

- **On-Premises:** All systems are stored and operated at your physical location(s). They may be independent servers and devices (like traditional networks) or they may be implemented as virtual machines in a private cloud. Either way, you have full control of the hardware on which they run and the location where they operate.
- **Cloud:** All systems are stored and operated at a cloud provider's physical location(s), such as AWS, Azure, or Google cloud services. You have no control over the specific hardware on which they run or even the exact physical location where they operate (though you can choose general regions in most cloud services and indicate that you desire hardware to be dedicated to your solution). However, from a logical operations perspective there is no difference. You can

---

[110] Interestingly, scalability can be considered from the workload perspective. That is, we can say that the workload can increase or decrease throughout the cycle of operations. A system that can handle a scaling workload must be powerful enough to handle the most intensive scale. This can result un underutilized resources for much of the operation time and is where cloud technologies shine, even private cloud solutions. Given that the systems run in virtual machines within the cloud, during times of the least intensive scale of a workload, the physical resources are available for improved performance in other systems running on the same physical hardware.

[111] Merriam-Webster Dictionary

use the same tools and processes to manage the system in the cloud as you would on-premises. An important possible difference, however, is latency. While it may vary only by a few or a few hundred milliseconds, latency is increased due to the need for traversal across the Internet.
- **Hybrid:** A hybrid cloud uses a public and private cloud together to implement a system. This should not be confused with multi-cloud, which simply indicates that your organization uses multiple clouds – some private and some public in many cases. Hybrid cloud indicates that the system depends on both clouds either continually or during specific operations (with dynamic cloud provisioning). Another view of hybrid implementations is the use of cloud and non-cloud solutions. You may have a single hardware-based server that runs a portion of the solution and use the cloud for the rest.

The selected architecture will also determine several requirements that must be defined:

- **Internet connectivity:** What throughput will be required of the Internet connection, if cloud services are used?
- **Local hardware:** What servers will be required locally for private cloud operations or traditional server systems?
- **Cloud services:** What service provider will be used, and do they offer IoT services that can assist in your solution?
- **Private cloud platform:** What private cloud platform will you deploy, it a private cloud is used, or do you already have one?

Scalability and elasticity can be essential components in IoT deployments. It is not uncommon to implement a portion of the IoT solution on year, another portion the next, and so on. Rather than building entirely new systems at each stage, existing systems can be scaled to accommodate new end devices, new data sets, and new analysis and visualization.

Coexistence

As the IoT design professional, you need to ensure the IoT systems coexist with other required systems within the ecosystem. Understanding the environment in which the installation will exist and the other systems that are in place is important. Coexistence is important from both a wireless and systems standpoint.

From a wireless standpoint, coexistence means utilizing an unused frequency or utilizing a technology that works well with existing wireless systems in the same frequency band.

For example, Zigbee and Bluetooth are both designed to operate in the 2.4 GHz band and work side-by-side with Wi-Fi. Within the 2.4 GHz band there are a lot of devices that can create interference with no desire to co-exist. Some of these include wireless analog cameras and cordless phones, microwaves, motion sensors, and more. In some environments the location will have a frequency coordinator and in others you may need to rely on data you capture.

From a systems standpoint, you need to make sure you can either interface into an existing system or provide a system that will work with what the customer has. In many cases, using a web-based service is a safe call as most users have Internet access and a web browser. This is not always the case, some IoT networks will be wholly contained as a private network or may be lacking backhaul to the Internet. In these cases, you must consider how the IoT solution can be accessed from existing systems to coexist with them.

4.4: Documenting the Requirements

We have discussed many considerations for requirements engineering in chapters 3 and 4. In this final section, we will make some recommendations for documenting the requirements developed. First, we will recommend several documents that should be created along the journey and then recommend possible solutions to common problems and conflicts that arise during requirements engineering.

Remember, conflict is normal in an organization. An organization is a collection of humans working together to achieve a goal. Conflicts can arise due to misconceptions about the organizational goals themselves, differences in opinion on how to reach the goals, personality driven conflicts, and misunderstanding of technology capabilities and uses, among other sources. Whatever the source, research has shown that documenting requirements helps alleviate conflicts and it has also shown common conflicts that arise, which will be addressed here.

Documents for Each Phase of Requirements Engineering

In chapter 3, you learned about business and stakeholder requirements. In this chapter, you learned about technical requirements. All three requirement types can be generated through individual but interrelated processes. In each of the three processes, documentation should be created to define the results. The 29148-2018 standard defines these three documents as Business Requirements Specification (BRS), Stakeholder Requirements Specification (StRS), and System Requirements Specification (SyRS). A fourth specification is also defined as Software Requirements Specification (SRS). We will not cover the latter in any level of detail. As a wireless IoT solution designer, you may request an SRS from developers working with you on the project but are less likely to be the manager of those requirements.

Business Requirements Specification (BRS)

The purpose of the BRS is to document and describe the organizations motivation for why the system is being developed. It defines processes and rules under which the system is used and documents the top-level requirements including the needs of users, operators, and maintainers based on the context of use of the system. It should be specific and unambiguous. While this document is defined separate from the stakeholder requirement specification in 29148-2018, in many real-world practices, the business and stakeholder requirements are simply called *business requirements* and stored in a single BRS or StRS document.

The following is an example outline provided in the standard for the BRS document[112]:

1. Introduction
    a. Business purpose
    b. Business scope
    c. Business overview
    d. Definitions
    e. Major stakeholders
2. References
3. Business management requirements
    a. Business environment
    b. Mission, goals, and objectives
    c. Business model
    d. Information environment
4. Business operational requirements
    a. Business processes

---

[112] ReqView is an excellent software program that maps to the 29148-2018 standard templates for the creation of the BRS, StRS, and SyRS documents. It is licensed on a subscription basis.

       b. Business operational policies and rules
       c. Business operational constraints
       d. Business operational modes
       e. Business operational quality
       f. Business structure
   5. Preliminary operational concept of proposed system
       a. Preliminary operational concept
       b. Preliminary operational scenarios
   6. Other preliminary life-cycle concepts
       a. Preliminary acquisition concept
       b. Preliminary deployment concept
       c. Preliminary support concept
       d. Preliminary retirement concept
   7. Project constraints
   8. Appendix
       a. Acronyms, definitions, and abbreviations

For guidance on the specific content in each of the sections above and those details for StRS and SyRS later, see the IEEE 15289-2019 standard which defines documentation procedures in system life cycle management.

Stakeholder Requirements Specification (StRS)
The StRS describes the motivation for the system being developed or changes from the stakeholders' perspective. It defines the needs of users, operators and maintainers from their active use perspective rather than the business-level perspective of the previous document. However, the StRS and BRS are often combined into a single document defining business requirements. If an StRS is created separately or a separate section within the business requirements document is used to define the stakeholder requirements, such requirements should be defined in close partnership with the stakeholders. The information elements within the document or section should be specified by the stakeholders and written for the purpose of documenting their needs and translating them into requirements.

The following is an example outline provided in the standard for the StRS document:

   1. Introduction
       a. Stakeholder purpose

b. Stakeholder scope
   c. Overview
   d. Definitions
   e. Stakeholders
2. References
3. Business management requirements
   a. Business environment
   b. Mission, goals, and objective
   c. Business model
   d. Information environment
4. System operational requirements
   a. System processes
   b. System operational policies and rules
   c. System operational constraints
   d. System operational modes and states
5. User requirements
6. Detailed live-cycle concepts or proposed system
   a. Operational concept
   b. Operational scenarios
   c. Acquisition concept
   d. Deployment concept
   e. Support concept
   f. Retirement concept
7. Project constraints
8. Appendix
   a. Acronyms, definitions, and abbreviations

System Requirements Specification (SyRS)

For the wireless IoT solution designer, if the processes recommended in 29148-2018 were followed, the SyRS becomes the primary guiding document during the design process. This fact is true because the SyRS defines the technical requirements that are based on the StRS and BRS before them. Therefore, if properly documented, the SrRS should provide the boundaries, capabilities, qualities, and other characteristics of the system that will ultimately accomplish the requirements in the StRS and BRS.

The standard states that the purpose of the SyRS is to provide a description of what the system should do in terms of the system's interactions or interfaces with its external environment. It should completely describe all inputs, outputs, and required relationships between inputs and outputs. Effectively, the SyRS communicates the requirements of the acquire and other stakeholders to the technical community that will design (specify) and build the system.

The following is an example outline provided in the standard for the SyRS document:

1. Introduction
    a. System purpose
    b. System scope
    c. System overview
        i. System context
        ii. System functions
        iii. User characteristics
2. References
3. System requirements
    a. Functional requirements
    b. Usability requirements
    c. Performance requirements
    d. Interface requirements
        i. External interfaces
        ii. Internal interfaces
    e. System operations
    f. System modes and states[113]

---

[113] A *system mode* may be defined as a characterization of the functional abilities of a system in a particular operational configuration in a particular context. For example, an IoT gateway may be in join mode to allow new devices to join the network, or it may be out of join mode when devices cannot join the network. System mode is generally considered as linked to a system state from the perspective of what the state allows. A *system state* is the condition a system is in based on the configurable parameters of the system and the processes it is running. System states may be linked to a system mode, but the system state is generally considered more variable than modes. There is significant disagreement in the requirements engineering community on the differences between system modes and system states or if they are simply synonymous. You will not be tested on a certain definition of these phrases but should understand that a system can be in a mode or state based on the configured parameters of the system and/or the stage in a process.

g. Physical characteristics
   h. Environmental conditions
   i. Security requirements
   j. Information management requirements
   k. Policy and regulation requirements
   l. System life cycle sustainment requirements
   m. Packaging, handling, shipping, and transportation requirements
4. Verification
5. Appendices
   a. Assumptions and dependencies
   b. Acronyms, definitions, and abbreviations

Resolving Problems and Conflicts

During the requirements engineering process many problems and conflicts will arise. This issue has been researched and analyzed extensively in the research literature and some of the most common issues that occur are addressed in this section. We intend only to provide suggestions on how to address such problems or conflicts and this section is highly opinion-based; therefore, it is provided for your review, but will not be tested on the CWIDP exam. We will address three issues based on statements that often surface them in the requirements engineering process.

Why should we invest the time required to perform requirements engineering?
This problem often surfaces very early in the process and is often an objection from management or the acquirer. When they discover that requirements engineering for a complex IoT solution can take weeks or months, they are often concerned about project delays and added costs. We suggest the following points as methods for addressing this concern:

- **Without proper requirements engineering, the total project time will typically take longer than it would with requirements engineering.** This reality is based on the work that will be repeated because the initial work does not meet the unknown needs of stakeholders. In the end, significant percentages of work performed was not even required to complete the project and the end result is far more time in the deployment phase than would have been necessary had requirements engineering been performed.

- **Without proper requirements engineering, project success cannot be measured.** To measure project success, the project must be properly defined, and this is what requirements do. Therefore, we can only say that we implemented some IoT solution at the end of the project if requirements engineering is not performed. However, if requirements engineering is performed, we can say that the right IoT solution was implemented.
- **Without proper requirements engineering, the users do not gain the commitment to the solution that they would had they been involved in the process.** This factor is very important to the success of any project: gaining user acceptance. When stakeholders have a voice in the requirements engineering process, they are far more likely to accept and use the resulting IoT solution.

But, I want this feature?

This conflict usually surfaces during stakeholder requirements planning. It is often considered a conflict between two requirements. One stakeholder wants one feature and another wants an additional feature that results in a conflict scenario because no single solution can implement both features or both features cannot be implemented within the budget constraints.

- **Elicit more information to arbitrate the conflict.** In many cases, discovering "why" the stakeholder desired a given feature will reveal that the feature is not actually required, and the need can be fulfilled by another non-conflicting feature of the solution. Remember, you are the technology expert and understand the needs that can be fulfilled by different system capabilities. The stakeholder simply knows what is desired to complete the job function. In many cases, no real conflict exists at the needs level and the feature-level conflict is illusory.
- **Ask each stakeholder how important the feature is to the fulfillment of responsibilities.** Ideally, each stakeholder should be asked for this information independently so that the second response does not simply "one up" the first response. With this priority information you can better inform the acquirer of the conflict and gain a decision for moving forward.
- **Ask each stakeholder what it would cost in time or finances if the feature was not available.** With this information, you can approach the acquirer to seek resolution. Having the priority and the cost factors can help the acquirer to determine if an increased budget is acceptable or if one or both features should be removed from the project requirements.

I thought we were...

This problem usually arises during the validation and optimization phase of a wireless IoT design process, but it is directly related to requirements. Two failures can result in the stakeholders thinking a particular result was to be achieved when it was not. The first is ineffective requirements engineering as a process. The second is poorly written requirement statements that are ambiguous or in some other way unclear. There is no need to extend into further explanations on how to prevent this because the entirety of the last two chapters has been focused on exactly that: performing effective requirements engineering and writing clear requirement statements.

4.5: Chapter Summary

In this chapter, you learned how to properly engineer technical requirements. Which critical technical questions to ask, which systems to think about when forming requirements, and how to link technical requirements back to the corresponding business requirements. You also learned the importance of the circular nature of requirements engineering to ensure requirements are feasible and complete. Keep in mind the questions you have learned to ask in this section as you learn the underlying different technologies so that you will know when each of these technologies will add value to the overall wireless IoT solution.

4.6: Review Questions

1. What is another phrase used to describe supporting services that might be used by a system of interest?

    a. System element

    b. Enabling systems

    c. System requirement

    d. None of these

2. True or False: A system element is one of the components that makes up a system.

    a. True

    b. False

3. True or False: An IoT system is often a system of systems including several different systems that work together to form the entire solution.

    a. True

    b. False

4. When is the functional boundary of a system defined within the 29148-2018 requirements engineering system requirements process?

    a. During preparation for system requirements definition

    b. During system requirements definition

    c. During system requirements analysis

    d. During system requirements management

5. What best defines resiliency?

    a. Using two routers in a subnet

    b. A measure of the ability of a system to continue operations in the presence of failures

    c. A measure of the number of servers used in a cluster

    d. None of these

6. What kind of interface is used between an end device and a gateway?

    a. System interface

    b. Connection interface

    c. User interface

    d. None of these

7. When two system elements within a system have interfaces, what is the common term for these interfaces?

    a. Internal

    b. External

    c. Accidental

    d. Incidental

8. What process is used to retrieve data from a source location, modify it if required, and store it in another location?

    a. BRS

    b. ETL

    c. SRS

    d. SQL

9. Which one of the following is an example of a NoSQL database?

    a. Oracle

    b. MySQL

    c. Cassandra

    d. MariaDB

10. What tool is used to present data analysis visually?

    a. PostgreSQL

    b. Grafana

    c. Firestore

    d. ReqView

4.7: Review Answers

1. The correct answer is B. Enabling systems reference supporting systems or services that provide services to the system of insterest.

2. The correct answer is A. A system is comprised of system elements. A system element may, in some cases, be another system.

3. The correct answer is A. An IoT system is often comprised of multiple systems and is called a system of systems.

4. The correct answer is A. The functional boundary of a system is defined during preparation for system requirements definition. The boundary constrains what must be defined in the requirements.

5. The correct answer is B. Resiliency is a measure of the ability of a system to continue operations in the presence of failures within the system. It is usually achieved with multiple areas of redundancy within the system elements.

6. The correct answer is B. A connection interface is used between end devices and the gateway.

7. The correct answer is A. Interfaces within a system are internal. Interfaces between systems are external.

8. The correct answer is B. Extract, transform, and load (ETL) is the process defined.

9. The correct answer is C. Of the databases listed, only Cassandra is a NoSQL database.

10. The correct answer is D. Grafana is a visualization tool that is used to create charts, graphs, and dashboards from analyzed data.

Chapter 5: IoT Architectures, Topologies, and Protocols

Objectives Covered:

3.1 Design for selected topologies

2.2.4 Gather and define system requirements (Wireless IoT architecture)

As a wireless IoT designer, you should understand the common IoT architectures, which define the systems, connections, and capabilities required in an IoT solution from end devices to business applications. The concepts include both architectural reference frameworks and concrete architectures. Creating an architecture is typically the first step between requirements engineering and the beginning of the design process or within the design process.

Additionally, you must understand the topologies used in various wireless IoT protocols. The topologies of the protocols determine, in part, the use cases for which they are most beneficial.

Both topics are explored in this chapter as well as an overview of common wireless IoT protocols. We will begin by clearly defining architecture and topology.

5.1: Architecture and Topology Defined

Having a shared meaning for terminology is important. So, we will begin this chapter by explaining our view on architecture and topology. This view is based on a common understanding in the networking community and an application to IoT specifically.

Architecture Defined

The concept of architecture comes from the world of building construction. The word architect is from ancient origins. From the oldest form in Greek[114], it was αρχιτεκτων (arkhitekton), which was from αρχι (arkhi, for chief) and τεκτων (tekton, for builder). From later Latin, it was architectus. From the 1500s in French, it was architecte. These terms were used to reference the individual who was a building designer or director of the building process. Over the years, the term architect has been prepended with many specifications to apply it to different industries: landscape architect, lighting architect, building architect, highway architect, and research architect, just to name a few.

---

[114] The concept of an architect dates back thousands of years. It is one of the most ancient occupations in the world, likely preceded only by farmer, warrior, toolmaker, and politicians.

Within the IT space, we have many architects as well: enterprise architect, solution architect, technical architect, network architect, system architect, and security architect, just to name a few. The goal of these roles is to define the architecture for the named entity or concept. An enterprise architect defines the enterprise architecture, and the network architect defines the network architecture. The architecture is a high-level design of a system and the logical and physical interrelationships between its components. From an IT perspective, architecture defines what is required to make an enterprise, solution, system, or system element work and the relationships with other enterprises, solutions, systems, or system elements that are required.

Architectures can be defined at a very high level with a simple diagram or at a very detailed level. However, when you begin to define specific APIs, functions, protocols, and applications, you are leaving the realm of architecture and entering the realm of system design. Think of the architecture as the logical or conceptual description and the design as the technical or practical description. Figure 5.1 shows a high-level architecture of an IoT solution. Notice that it does not explain the links between the gateway and the servers in any detail. It does not define the specific Internet link type to the cloud from the gateway or the data aggregator. It simply defines the components and their relationships.

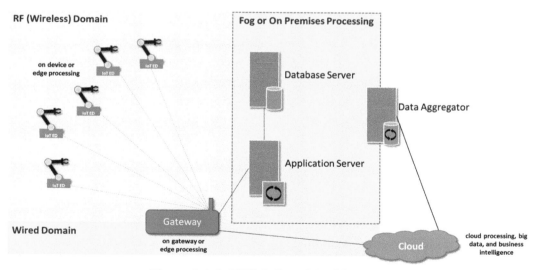

Figure 5.1: IoT High-Level Architecture

An example architecture from the wired networking domain is the popular Core, Distribution, Access model. It provides device access to the network at the Access layer, uses the Distribution layer for delivery among Access layer devices, when optimum, and uses the Core for high-speed communications across the backbone of the network. It is represented in Figure 5.2.

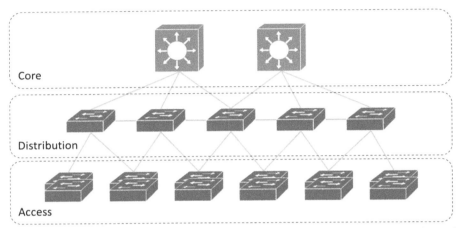

Figure 5.2: Core, Distribution, and Access Layer Network Architectural Model

You can also divide architectures into reference frameworks and concrete. An architectural reference framework is a reference model used to create concrete architectures. That is, the framework identifies common architectural components used in building a solution of a given type and the concrete frameworks defines the components used in a specific implementation of that solution type. Later sections will explore wireless IoT architectural reference frameworks and concrete architectures in more detail.

Finally, when building an architecture for an IoT solution, the primary goal is to develop a reference point that can be used during the rest of the design process. The architecture should be derived from the requirements and the design can be derived from the architecture (additionally, a reference architecture is a good source for input to the requirements engineering process). However, you can view an architecture from different views or perspectives. From this concept was born the ideas of *architectural views* and *perspectives*. Architectural views focus on system features that can be isolated and make the diagrams and concepts easier to understand, particularly for non-technical stakeholders. Common architectural views include:

- **Functional:** Comprises the functional-decomposition viewpoint[115] (the breakdown of separate functions or function groups), the interaction viewpoint (the relations between functions and other components), and the interface viewpoint (the methods of data request and transfer between components or other systems).
- **Information:** Comprises the information flow viewpoint, information processing viewpoint, and information hierarchy viewpoint.
- **Deployment or Operational:** Comprises the application viewpoint, the services viewpoint, the network connectivity viewpoint, and the devices viewpoint. This is the most common view a wireless IoT designer would utilize and is what most wireless IoT designers simply call the IoT architecture or concrete IoT architecture[116].

Architectural perspectives assist in defining those concepts that are cross-cutting in nature. That is, they span across the various architectural views. These concepts are usually qualitative in nature and are driven by quality requirements, which we discussed earlier in the book. In fact, an architectural perspective has been defined as *a collection of activities, tactics, and guidelines that are used to ensure that a system exhibits a particular set of related quality properties that require consideration across a number of the system's architectural views*[117]. They further define a *quality property* as *an externally visible, non-functional property of a system such as performance, security, or scalability*. So, we are back to the "ilities" again and quality requirements.

In most cases, architectural perspectives are used as an analysis function in relation to the architecture. The designer analysis the security, performance, or other architectural perspective in relation to the proposed architecture and makes changes as required. The IoT-ARM provides excellent tables of activities and tactics to assist in this process. The

---

[115] A *viewpoint* can be defined as a collection of patterns, methods, and best practices for building a type of architectural view.

[116] My statement that the Deployment and Operational Architectural View is the most common is not intended to mean it is the best. In general, I recommend that more views be used on the largest and most complex of projects down to the fewest views on the smallest and least complex.

[117] Rozanski and Woods. *Software Systems Architecture*. 2005. www.viewpoints-and-perspectives.info/home/perspectives/

group that developed the IoT-A has been decommissioned as they have completed their task; however, the model is still one of the most commonly referenced models for IoT architectures and the document is still available at archive.org[118]. Table 5.1 is based on the Performance and Scalability perspective as documented in the IoT-A document.

Desired Quality	The ability of the system to predictably execute within its mandated performance profile and to handle increased processing volumes in the future, if required.
Applicability	Any system with complex, unclear, or ambitious performance requirements; systems whose architecture includes elements whose performance it unknown; and systems where future expansion is likely to be significant. IoT systems are very likely to have unclear performance characteristics, due to their heterogeneity and high connectivity of devices.
Activities	Capture performance requirements; Create performance models; Analyze the performance model; Conduct practical testing; Assess against the requirements; Rework the architecture
Tactics	Optimize repeated processing; Reduce contention via replication; Prioritize processing; Consolidate related workloads; Distribute processing over time; Minimize the use of shared resources; Reuse resources and results; Partition and parallelize; Scale up or scale out; Degrade gracefully; Use asynchronous processing; Relax transactional consistency; Make design compromises

Table 5.1: Performance and Scalability (Enhanced from IoT-A Documentation)

Topology Defined

The term *topology* like the term architect, originates in the Greek from the word τοπος (topos). The word meant place or location in the Greek. Sometime in the 15th century it became common use with the suffix *ology* as topology to reference the place or locations where certain plants are found. Over time it evolved into technical use in the sciences and eventually made its way into the computer networking world in the 1970s as *network topology*. You can think of it as the placement (from τοποσ) of nodes in relation to one another.

---

[118] web.archive.org/web/20150118071217/http://www.iot-a.eu/public/public-documents/d1.5/at_download/file

In a December 1978 thesis by Donald L Ravenscroft entitled *Electrical Engineering Digital Design Laboratory Communications Network*, the author presented four basic topologies as shown in Figure 5.3. What the author calls a *loop* topology eventually became known as a ring topology in most systems. The star and tree topologies still exist today with little difference in them. The *distributed* topology is referenced today as a mesh or partial mesh topology. Intrinsic to all the topologies is the most fundamental link type: point-to-point (PtP). PtP links must exist, and in many cases multiple of them, to form the various topologies.

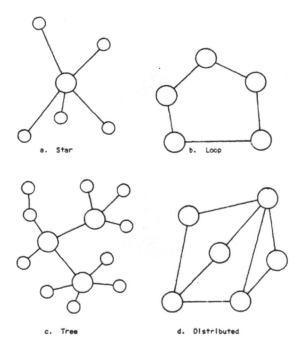

Figure 5.3: Four Basic Network Topologies from a 1978 Thesis Paper

The topology defines how the network components are connected with each other. For example, all components may be connected to one central component, or they may all have connections with each other as well or instead. Depending on the interconnections provided, the topologies use different terms to summarize them.

Much of our terminology and conceptualization comes from the mathematical specialization of graph theory (or network graph theory), which is the study of graphs. Not the graphs that we create in Excel or PowerPoint, but the graphs consisting of multiple objects organized into a structure where some objects are related to each other. In graph theory, topological properties are discussed, which reveal the sub-structures within a network (a graph network). Additionally, in graph theory, the topology is the way in which the nodes and edges are arranged in the network[119].

In keeping with the realm from which the terms are borrowed, communication network topology is the definition of the nodes and their connections with one another. Differentiating this from the system architecture allows us to see the big picture and also zoom in on the details as required.

The challenging thing is the non-technical way in which both the term architecture and topology are used. It is not uncommon to see references to network topology architectures, which blurs the two concepts. For our purposes, the topology will define how the network components or devices connect to other components or devices and the architecture will picture the entire network from a logical point of view.

5.2: Wireless IoT Architectural Reference Frameworks

As stated previously, an architectural reference framework is not intended to be the architecture implemented in a given system but is intended to act as a guide for the creation of a concrete architecture – one that will be implemented in a given system. The existing IoT architectural reference frameworks (or sometimes just IoT reference architectures) are based on the assumption that IoT solutions can be abstracted to such a point where the most commonly used components, characteristics, and requirements can be identified within the model[120]. Then, engineers can use the model to determine that

---

[119] While the discussion of network graph theory could continue to topics like degree, shortest path, transitivity, centralities, topological clusters, and more. You can see from what is discussed, that the terms we use in communications networking come from this much older mathematical world of topology, graph theory, and other areas of mathematics dealing with graphs (in some cases dating back hundreds of years).

[120] And it is true that the IoT solutions can be abstracted; however, be forewarned that this does not mean such reference architectures are not complex.

which should exist within the concrete architecture (the real architecture to be implemented) in their projects.

Several such reference models exist, and this section will explore those that are most commonly referenced, no pun intended, and explain how they might apply. We will begin with the simplest model and expand to the more complex[121].

The Three-Layer Model

The three-layer IoT reference model is the simplest. It is also a good starting point for discussions very early in IoT projects. This model is represented in Figure 5.4. In the image, the bar across the top represents the three layers: Perception, Network, and Application. In this model, the Perception layer is where the sensors, actuators, and things reside. The Network layer allows for connectivity of the Perception layer devices and transmission of data to the Application layer. The Application layer is all inclusive of data collection, initial storage, processing, distributed storage, and APIs. The purpose of the Application layer services is to provide business decisions, insights, and possibly control of the things within the Perception layer.

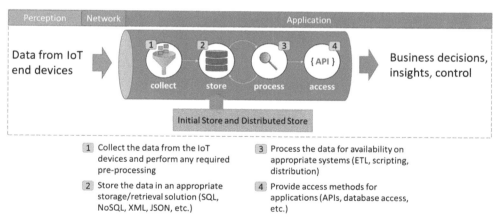

Figure 5.4: The Three-Layer Model of IoT

---

[121] While many resources use the phrase *reference architecture* to refer to the final architecture, we are referring to that as the *concrete architecture* to avoid confusion with the phrase *reference architecture framework*.

The three-layer model is perhaps most useful when communicating with business decision makers and stakeholders without a technical depth of understanding. It maintains a level of simplicity that allows for early-stage conceptual models that can later be expanded to more complex architectures for the final system.

Expanded Three-Layer Models

The next two models are expansions on the three-layer model by extracting portions of the Application layer into one or two additional layers. The first is a four-layer model and the second is a five-layer model. All three models are represented in Figure 5.5. The green boxes represent the end devices. The grey boxes are roughly equal to the OSI model, and the blue boxes are described below.

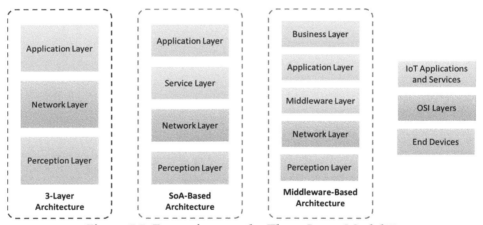

Figure 5.5: Expansions on the Three-Layer Model[122]

The Service-Oriented Architecture (SoA)-Based architectural model extracts the services, such as data collection, storage, and processing out of the Application layer into the Service layer forming a four-layer model. This model allows for discussion of the services separate from the applications and allows the services to provide functionality to various different applications based on the same lower layers. The Application layer, in this model, is still inclusive of APIs and other network-based business logic and the business

---

[122] Lombardi, Pascale, and Santaniello, Internet of Things: A General Overview between Architectures, Protocols and Applications, 2021, Information journal

applications that will access them. In the SoA-Based Architecture model, the service layer is responsible for service discovery, service provisioning, service management, and interfaces to the services.

The Middleware-Based Architecture model, or the five-layer model, uses a Middleware layer between the Network and Application layer and adds a Business layer at the top. The Middleware layer provides for multiple kinds of communications from the network to work with multiple applications. Middleware is software that exists between an operating system and the applications running on it, essentially functioning as a translation layer. It enables communication and data management for distributed applications[123]. It provides common services and capabilities to the applications outside of what's offered by the operating system[124]. Because middleware is such an all-encompassing concept, it can include APIs, message brokers, abstraction layers, and authentication system. Fog, edge, and cloud computing are all sometimes considered as middleware layers as well as Big Data storage and analytics and other information processing systems and functions[125].

In the five-layer model, the Applications layer is also divided further into an Application and Business layer. The Application layer provides the various tools and frameworks that are used for smart analysis, decision making, and management. The Business layer adds in overall system monitoring and management and data visualization functions with smart graphs, charts, and statistics views.

We can see from these expanded models how the basic three-layer architecture reference model can be expanded easily to four or five layers and provide more details. As you can imagine, we could further divide the layers and develop other architecture models. Consider, for example, Figure 5.6 where the Perception Layer is expanded to include Things, Sensors, Actuators, Compute Hardware, and Radio Modules. The Network Layer is expanded to include a Physical Layer, Data Link Layer, Network Layer and Transport Layer. This could be expanded further, for example, so that the Data Link Layer is broken into components such as security, data delivery, and link management. The Network

---

[123] *What is middleware?* azure.microsoft.com/en-in/overview/what-is-middleware
[124] *What is middleware?* redhat.com/en/topics/middleware/what-it-middleware
[125] Isravel, Deva Priya. *A Novel Framework for Quality Care in Assisting Chronically Impaired Patients with Ubiquitous Computing and Ambient Intelligent Technologies*. Systems Simulation and Modeling for Cloud Computing and Big Data Applications. 2020.

Layer could be broken into addressing, fragmentation, compression, security, and reliable delivery, and more. The point of a reference architecture for a specific technology, such as IoT, is to provide enough detail to be applicable to all systems but insufficient detail to rule out particular systems. A concrete architecture moves to the next level of ruling out some or most systems as it defines a unique system.

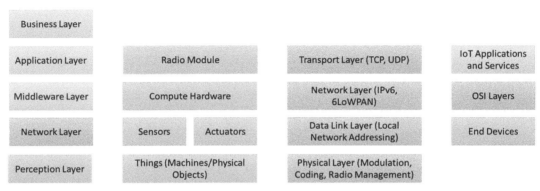

Figure 5.6: Multiple Expansions of Architectural Layers

Another way of conceptualizing the architecture is with the simple division of Network and Applications. This is shown in Figure 5.7. It is based on the system of systems thinking model wherein the layers can be further subdivided into ever more layers.

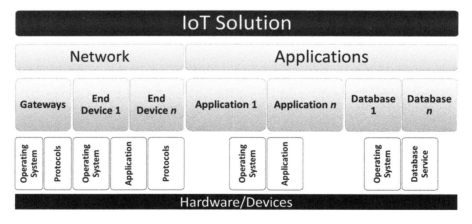

Figure 5.7: System of Systems Architecture Model

Finally, before moving on to more complex reference architectures, consider the use of even the simplest three-layer model in the development of an IoT taxonomy[126], which is useful as a reference in planning and design and communications. Figure 5.8 illustrates such a taxonomy.

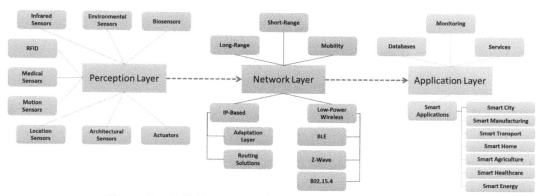

Figure 5.8: IoT Taxonomy from the Three-Layer Model

The IoT-A Reference Model

The IoT-Architecture Reference Model (sometimes called IoT-ARM or simply IoT-A) was one of the earliest[127]. It was completed by a special commission in 2013 and resulted in a nearly 500-page document describing an architecture reference model and the components that comprise it. Even though it is one of the earlier models and is no longer being directly updated, it is still one of the most comprehensive of the models.

The IoT-A document begins with a very high-level overview of the IoT-A architecture reference model with the metaphor of a tree (depicted in Figure 5.9). The root of the tree is the devices and communication protocols, and the leaves of the tree are the

---

[126] A *taxonomy* is a system of classification that is often hierarchical. That presented here is an architecture-based taxonomy. Such a taxonomy can be created for a concrete architecture that you develop from an architectural framework as well. Once created, it acts as a reference to the many components or elements of the IoT solution.

[127] You will not be tested on the specific details of a given IoT reference architecture, but you should know how one is used in the creation of a concrete architecture.

applications and use cases. The trunk of the tree represents the information and knowledge made available by and to the IoT devices.

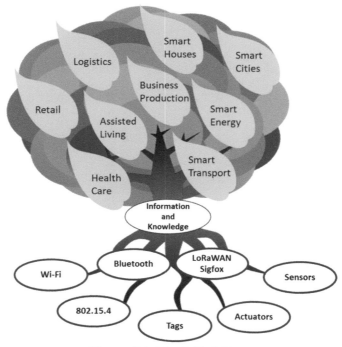

Figure 5.9: The IoT-A Tree

IoT-A divides the reference into an IoT Reference Model and an IoT Reference Architecture. The IoT Reference Model consists of multiple sub-models beginning with the IoT Domain Model as the foundation and including the following models:

- IoT Domain Model
- IoT Information Model
- IoT Functional Model
- IoT Communication Model (Functional Group of the Functional Model)
- IoT Trust, Security, and Privacy Model (Functional Group of the Functional Model)

These models and their interrelationships are depicted in Figure 5.10. The IoT Domain Model presents concepts that are modelled and represented in IoT systems and provides input to the IoT Information Model. The IoT Domain Model also presents concepts as foundations of Functional Groups and provides input to the IoT Functional Model. In addition, the IoT Information Model provides inputs to the IoT Functional Model by providing information elements to be handled by the Functional Groups (FGs). All of this is illustrated in Figure 5.10. The FGs shown in Figure 5.10 are just two of the possible FGs.

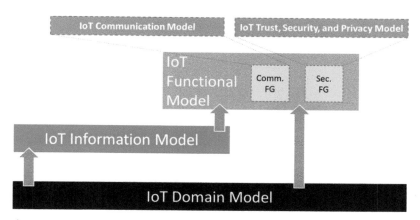

Figure 5.10: IoT-A Reference Model, Sub-Models, and Interactions

IoT Domain Model

A domain model[128] is a description of the concepts contained within an area of interest (the domain). It defines the properties of the concepts and the relationships between them. Properties might include names and identifiers. Relationships include inputs to and outputs from one concept to or from another. The goal of the IoT Domain Model is to define a common understanding of the IoT world.

Within the IoT-A documentation, the IoT Domain Model is represented in the Unified Modeling Language (UML). To avoid confusion for those unfamiliar with UML, we will present it here in a simple descriptive diagram. The model can be separated into three primary concept types. Human users (yellow), virtual users and devices (brown), which

---

[128] As a more thorough definition, Eric Evans defines the phrase *domain model* in *Domain-Driven Design Reference: Definitions and Pattern Summaries*, 2015, as *a system of abstractions that describes selected aspects of a sphere of knowledge, influence, or activity (domain) and can be used to solve problems related to that domain.*

can be humans represented within the system or other devices/systems, digital concepts (green), and physical devices (blue). The model is shown in Figure 5.11[129].

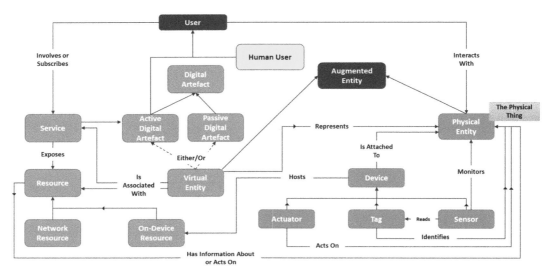

Figure 5.11: The IoT Domain Model

The following will explain the concepts in the IoT Domain Model:

- **Physical Entity (The Physical Thing):** This is the machine, environment, vehicle, or any *thing* that has an IoT device attached or is embedded with an IoT device for sensing, identification, or actuation. The Physical Entity is represented digitally within the IoT system as a Virtual Entity, which is explained later.
- **Device:** The component attached to or embedded in the Physical Entity that makes it a cyber physical object and provides for IoT connectivity and actions and may include one or all the following:
    - **Sensor:** A component that senses something about the physical object or environment for which it is deployed. This may include temperature, humidity, motion, location (with or without the use of a tag), speed, active/inactive, open/closed, fluid levels, flow rates, and many more.
    - **Tag:** In wireless IoT, an RF component that contains the ID of the attached device and may be used for location tracking in some systems.

---

[129] We have maintained the British spelling of Artefact used in the original IoT-A documentation.

Read by some sensing device, such as an RFID reader/scanner, if it is an RFID tag.
- **Actuator:** A component that can modify the physical state of the Physical Entity to which the device is attached or in which the device is embedded. For example, it may inflate, deflate, rotate, stir, activate, deactivate, increase, decrease, or any number of other actions for which it is configured to operate.
- **Virtual Entity:** This entity represents the Physical Entity in the digital space. It is what applications and services use to read parameters of the Physical Entity (provided by the Device) and send commands to the Device attached to or embedded in the Physical Entity. A Virtual Entity is a Digital Artefact that is either Active or Passive.
  - **Digital Artefact:** Generally speaking, in the context of use here, a digital artefact (or artifact) is any item produced and stored as a digital or electronic version. Therefore, Virtual Entities are Digital Artefacts of the Physical Entities. All Digital Artefacts are either active or passive.
    - **Active Digital Artefact (ADA):** An ADA runs a software application, agent or services that may access other services our resources.
    - **Passive Digital Artefact (PDA):** A PDA is a representation of a Physical Entity that performs no actions but can be inspected by the system.
- **Augmented Entity:** This is the composition of the Physical Entity and the Virtual Entity that are associated. From a system perspective, the applications and users need not consider if they are communicating with a Physical or Virtual Entity. They are, from their perspective, communicating with the Physical Entity, but the system may consider the Augmented Entity and the Virtual Entity through which the Physical Entity is accessed. The Augmented Entity exists in the model primarily to indicate that the Physical and Virtual Entities are associated.
- **Resource:** A software module or component that provides data from Physical Entities (or the Devices monitoring them) or that performs actuation on Physical Entities based on requests from users (either Human or Digital):
  - **On-Device Resource:** These resources are hosted by the device. It is software running on the device that can provide information to the network or that can perform actuation on the Physical Entity.

- **Network Resource:** These resources exist somewhere in the network.
- **Service:** A Service exposes a Resource or Resources. The Service is the software that presents the Virtual Entity to the network as a collection of one or more Resources.
- **User:** The User is a logical entity in the domain model that may represent a Human User or an Active Digital Artefact. That is, a User may be a human or another device or application on the network.

As you contemplate the modeled concepts and entities listed above and presented in Figure 5.11, you can see that much thought went into the development of this model, even though it is nearly ten years old at the time of this writing. Newer IoT reference architectures and models have presented additional conceptual factors, but the modeling of the IoT domain itself has rarely been presented as well as this.

With an understanding of the IoT Domain Model, you can better understand the other models within the IoT-A framework. Next, we'll investigate the IoT Information Model.

IoT Information Model

The IoT Information Model is simpler than the IoT Domain Model as it models the conceptual details of the information related to a Virtual Entity alone. The Virtual Entity, as seen in Figure 5.11 earlier, relates to a Service, which relates to a Resource (either Network or On-Device). Figure 5.12 shows, again without UML, the relationships in the IoT Information Model related to these concepts.

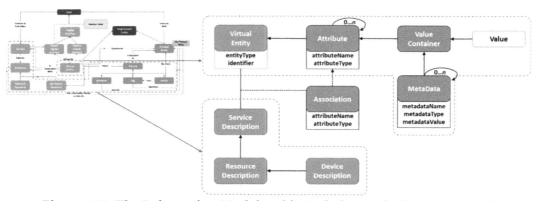

Figure 5.12: The Information Model and its Relation to the Domain Model

The IoT Information Model identifies how the attributes and values of a Virtual Entity are provided within the IoT system. First, a Virtual Entity has an *entityType* and an *identifier*. The *entityType* could be a human, trailer, boiler, environmental sensor, or any number of other IoT types. The *identifier* is simply a unique identifier for the entity.

Each Virtual Entity may have 0 to *n* (many) attributes and the attributes have the properties of *attributeName* and *attributeType*. A Virtual Entity may be defined in a separate ontology as having a minimum and optional set of attributes. The Attribute, in turn has a Value Container which holds a value and may have 0 to *n* MetaData elements associated with it. For example, the value may be 22 and represent a temperature. The MetaData could include a timestamp for when the temperature was reported, a unit of measurement (such as Celsius), and a degree of accuracy, among others.

So far, we've described the information *about* the Virtual Entity defined in the Information Model. Now, we will look at the Service that exposes this Virtual Entity information. The Service Description describes the aspects of the service including its interface. The Resource Description describes the Resource whose functionality is exposed by the service and the Resource Description may contain information about the Device on which the Resource is hosted.

Finally, the Association is the link between the Service that exposes the information about the Virtual Entity and the Attributes of that Virtual Entity. It is described as a *get* function used to retrieve information, but it can also be considered a *put* function when considering actuators or the possibility of configuration of Devices.

IoT Functional Model

The IoT-A IoT Functional Model is based on *Functional Decomposition (FD)*, which is the process by which the various functional components that make up the reference model are identified and related. The goal of FD is to simplify a complex system by defining its individual functional components rather than attempting to define the system as a whole. Through FD, the IoT-A reference model provides an abstract framework for understanding the main Functionality Groups (FGs) and their interactions. This framework defines the common semantics of the main functionalities and will be used for the development of IoT-A compliant functional views[130]. A functional view describes

---

[130] Internet of Things – Architecture (IoT-A) Deliverable 1.5, 2013

the functional components and their responsibilities, functions, interfaces, and primary interactions. Figure 5.13 shows the IoT-A Functional Model.

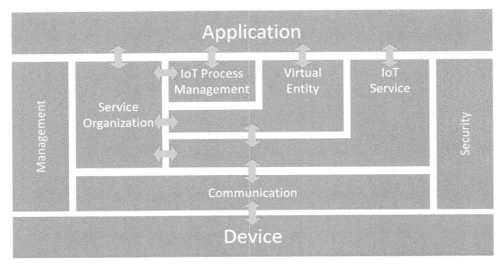

Figure 5.13: IoT-A Functional Model

The Functional Model defines nine different FGs. Seven are defined as longitudinal (blue in the image) in the IoT-A documentation and two are defined as transversal (orange in the image: Management and Security). These terms simply mean that the Management and Security FGs are providers for all the other FGs. For example, security is needed from the Device to the Application and everywhere in between, as is management. In Figure 5.13, the interactions between the longitudinal FGs are depicted by the yellow arrows. The FGs are describes as follows:

- **Device:** This is the Device as defined in the IoT Domain Model.
- **Communication:** This FG is focused on the network communications from the devices to the IoT Service. In wireless IoT, this could be 802.15.4, 802.11, Bluetooth, LoRaWAN, or any number of other protocols with security and management capabilities integrated within them.
- **IoT Service:** The collection of one or more IoT services required of the solution and functions for the discovery and name resolution of IoT services.

- **Virtual Entity:** This is the Virtual Entity as defined in the IoT Domain Model and IoT Information Model.
- **IoT Process Management:** The functions used to integrate the IoT solution with traditional business processes. This may include ETL operations, APIs, data access, and more.
- **Service Organization:** A hub and/or abstraction layer allowing the different FGs to communicate with each other or allowing external applications to communicate with the IoT solution. The Service Organization exposes the IoT Process Management, Virtual Entity, and IoT Services functions to other FGs and external applications.
- **Management:** The functions that are required to govern an IoT system including fault handling and cost management. Measures implemented for fault handling may include prediction of potential failures, detection of a failure, mitigating the impact of failures, and repair functions. Measures implement for cost management may include logging usage by a department or group for billing or managing the workload effectively to allow for more users/devices without increased cost.
- **Security:** The functions that provide for secure joining of the IoT system, trust between IoT devices, confidentiality of communications, integrity of communications, and other security requirements.

The IoT Communications Model and Security Model will not be discussed here. However, security will be addressed in other parts of the book. The Communications Model simply defines different ways to conceptualize communications related to IoT gateways and devices.

With this IoT-A Reference Model overview, we can explore the IoT Reference Architecture from IoT-A.

The IoT-A Reference Architecture

It is time now to answer the important question: How does this model that we have covered in the preceding section become a reference architecture? The answer is found in Figure 5.14. You will recognize the concepts reflected in the reference architecture shown (this is the functional view of the IoT-A reference architecture) and see that some specific functions have been added that are required to effectively implement the model.

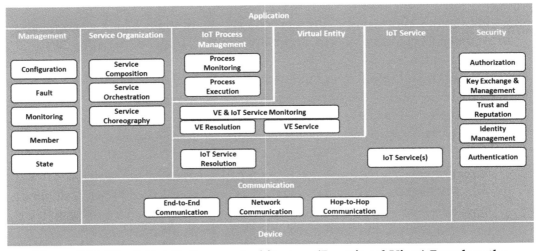

Figure 5.14: The IoT-A Reference Architecture (Functional View) Based on the Reference Model

The simplest way to conceptualize how this reference model works is to consider a few examples. First, Figure 5.15 removes anything not actively required for an example of a pressure sensor reporting to a mobile phone application based on a subscription to the reports. We are assuming, in this case, that the end device (the pressure sensor) and the mobile phone application have already onboarded to the system and have properly authenticated and derived encryption keys. So, the only thing remaining is to communicate. The following steps outline what is depicted in Figure 5.15:

1. The pressure sensor detects a pressure change and the device monitoring the sensor reports it to the communication FG in a payload targeted at the IoT Service.
2. The IoT service is reached through some number of hops between the end device and the service.
3. The IoT service records the pressure reading with the Virtual Entity (VE) service for application of the VE attributeType (*pressure*).
4. The mobile phone application, which has already subscribed to updates from this VE receives the pressure change reading and responds appropriately (alerts the users, does nothing, triggers an action, etc.).

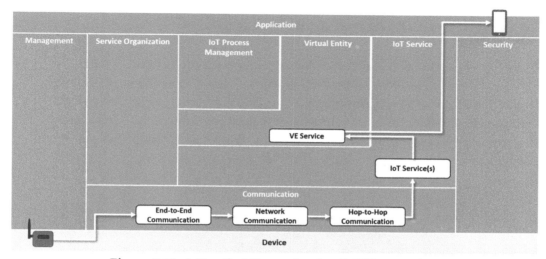

Figure 5.15: A Practical Example of an IoT Transaction

At this point the IoT-A reference model and architecture have been addressed sufficiently to see the value of such an architecture. Before exploring the process of creating a concrete architecture based on a reference architecture, we will very briefly explore two other reference architectures: The Industrial Internet of Things Reference Architecture and the oneM2M Reference Architecture, the latter being a "real" system that can be implemented or used as a reference.

Industrial Internet Consortium (IIC) IIoT Architecture Framework

The IIC has developed the Industrial IoT (IIoT) Reference Architecture (IIRA) for IIoT systems. The IIRA specifies the Industrial Internet Architecture Framework (IIAF) with viewpoints and concerns to assist engineers in the development, documentation, and communication of the IIRA. The IIRA is used by system architects and engineers as a guide or template for defining a specific IIoT system. Like the IoT-A reference architecture, the IIRA can be used to assist in requirements engineering and in the process of creating a concrete architecture. Using such reference architectures can assist in the consistent architectural implementations across varying use cases in the different industrial sectors so that unique system requirements may be met[131]. The reference architecture also provides a shared vocabulary and model for effective understanding

---

[131] *The Industrial Internet of Things Volume G1: Reference Architecture.* Version 1.8. 2017

and collaboration among engineers, designers, stakeholders, acquirers, and other involved parties.

Figure 5.16 shows the IIAF functional view. It begins with the Physical Systems (the things and devices), includes Functional Domains (Control, Operations, Information, Application, and Business), Crosscutting Functions (Connectivity, Distributed Data Management, Industrial Analytics, Intelligent & Resilient Control, etc.), and System Characteristics (Trustworthiness (Safety, Security, Resilience, Reliability, Privacy), Scalability, etc.).

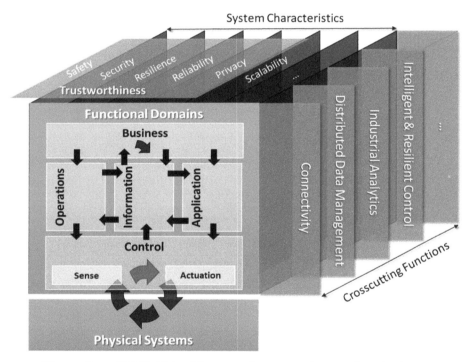

Figure 5.16: IIC IIoT Architecture Framework Functional View[132]

Within the IIRA documentation, an analogy is provided that makes the benefits of a reference architecture clear. It states:

---

[132] Reproduced from *The Industrial Internet of Things Volume G1: Reference Architecture*. Version 1.8. 2017

A reference architecture for a residential house states that all residential houses need to provide one or more bedrooms, bathrooms, a kitchen and a living area. This set of rooms is accessible inside the house through doors, hallways, and stairways, and from outside through a main and a back door. The house provides a safe environment against threats such as fires, hurricanes, and earthquakes. The structure of the house needs to sustain snow and wind load that may be found in its local environment. The house needs to provide reasonable measures to detect and prevent unauthorized intrusions[133].

Notice that the preceding analogy is specific enough to indicate several concepts required of a proper residential home; however, it is abstract enough to allow for one bedroom or ten bedrooms, one bathroom, or five bathrooms, two ingress/egress doors or eight, etc. The point is that a reference architecture defines what is most frequently required, but also allows for additions and removals as the system of interest demands.

We will not explore the IIRA or IIAF more here, but it can be freely downloaded, as well as many other related guidelines and models, from IIConsortium.org. The point here was only to briefly introduce a market-specific framework that can be considered and learned from by other markets as it is one of the best-defined frameworks available for any market[134].

oneM2M Functional Architecture

The final reference architecture we will review is the oneM2M architecture[135]. This model has been adopted by ETSI as a base architecture for there SmartM2M guidelines. For example, they adopt it without significant change in ETSI TS 118 101 3.22.0 oneM2M - Functional Architecture (2021). Then, the SmartM2M recommendations they provide are founded upon this architecture.

---

[133] Ibid. clause 2

[134] While the IIC IIRA is targeted at the IIoT market, the reality is that it can be customized easily to fit into any market. The basic concepts are universal and only those specific to IIoT would require removal or modification. If you work in a large organization that will deploy hundreds or thousands of wireless IoT solutions, consider starting with the reference frameworks presented in this chapter to develop a reference framework that will function well for your organization across all projects.

[135] oneM2M is a middleware services specification but acts as an excellent reference point for designing architectures that use oneM2M or that use other solutions.

oneM2M was launched in 2012 as a global initiative to develop specifications that ensure the most efficient deployment of Machine-to-Machine (M2M) communications systems and the IoT. They develop a common service layer that can be embedded within hardware and software to allow connectivity for varying devices and that can communicate with M2M compatible application servers. Involved organizations include Cisco, HPE (HP Enterprise), Huawei, Qualcomm, and others[136].

Figure 5.17 shows the oneM2M high-level functional architecture. It simply defines three entities that should exist within the Field Domain and the Infrastructure Domain. The Field Domain is where the things live, and the Infrastructure Domain is where the business or organizational applications and service live.

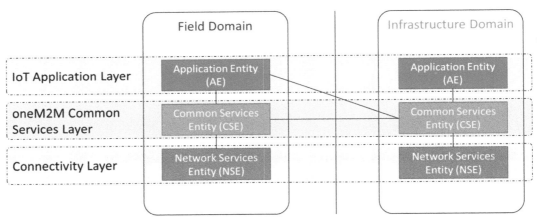

Figure 5.17: oneM2M Functional Architecture

The model is simple because each entity is described in significant detail so that a system can be built to be compatible with the oneM2M framework. This concept is where oneM2M is somewhat different from other frameworks. The Common Services Layer (CSL) is the core of the model. Each device (end devices, application servers, etc.) implements a Common Services Entity (CSE) that provides the functions of the CSL. It is primarily focused on developing devices and platforms that comply with the model.

---

[136] oneM2M.org

However, as a reference, it provides rich insights into the concepts and processes required to build an IoT architecture.

For example, the published specifications cover topics such as security, IoT Data Protocol integration, interoperability, and device configuration. The collection of published materials contains a wealth of information for the IoT designer and architect.

5.3: Creating a Concrete Architecture

As a wireless IoT designer, architecture reference models will become very important to you. Whatever reference model you adopt over time, whether those described in the preceding section or one of your own creation, this model will become the foundation one which you design and build IoT solutions. You can think of the reference model as the conceptual blueprint that you customize through profiles to invoke a concrete architecture for a given application. Figure 5.18 illustrates this concept.

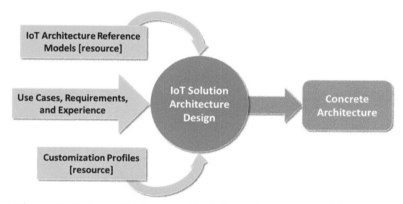

Figure 5.18: From Reference Models to Concrete Architectures

The *concrete architecture* is an instantiation of the reference model or framework applied to a practical use case[137]. Vendor architectures are concrete architectures (though sometimes they are closer to reference architectures that you customize during

---

[137] Bassi, Bauer, Fiedler, et. al., *Enabling Things to Talk: Designing IoT Solutions with the IoT Architectural Reference Model.* 2013

implementation) for the use of each vendor's solution. Custom concrete architectures are those you create for a specific scenario. A vendor or custom concrete architecture may define specific technologies used, though it will not define the details of configuration. Such details will be provided in the design.

Vendor Concrete Architectures

Vendors, often cloud service providers, provide architecture definitions that are summarized across seven vendors in Figure 5.19. While they vary in many cases, they have also settled upon some consistent components. For example, they all begin with the end devices and the connectivity of those devices through various gateways. In Figure 5.19, the color coding represents the Perception layer (purple), the Network layer (green), the Middleware or Services layer (blue), the Application layer (orange), and the Business layer (purple).

Vendor architectures can be useful to understand concepts common to architectures. Notice that several architectures address edge and fog computing. Other address data storage, data accumulation, and warm path vs. cold path store. These concepts provide essential ways of thinking about compute and data acquisition. Where will compute occur in your architecture? Where will data be stored? Will it be processed during ingress (streaming) and stored afterward? Will it be stored and then processed? By inspecting the architectures of others, you can better understand what should be included in your concrete architecture.

Of course, if you are using one of the referenced vendors, then understanding their architecture is essential. Be sure to check the vendor documentation as these architectural models change over time and those referenced in Figure 5.19 are simplified based on analysis of the primary components of the listed architectures by Ameyed, Petrillo, Jaafar, et. al.

AWS has some of the best documentation among the cloud service providers and they have a rich IoT offering for the data, analytics, machine learning, and application components of an IoT solution. To view documentation for their IoT Core solutions, visit docs.aws.amazon.com/iot/index.html. AWS IoT Core supports MQTT, MQTT over Websockets Secure (WSS), HTTPS (REST), and has integrated support for LoRaWAN directly (though any communications that ultimately traverse IP can reach the cloud).

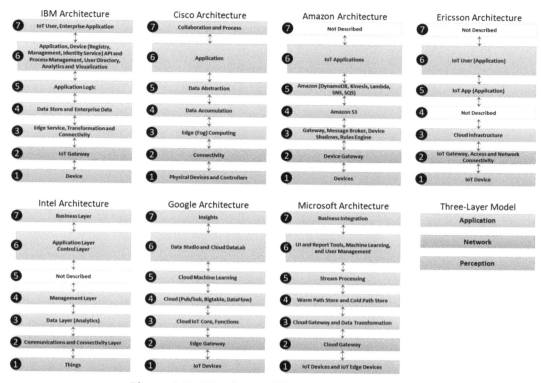

Figure 5.19: Vendor Architectures Compared[138]

Custom Concrete Architectures

Having explored reference architectures and models as well as vendor architectures, you can begin the process of creating your concrete architecture for a specific system of interest. Each IoT solution will have some unique qualities, and this is even true in an organization deploying the same basic IoT capabilities at different locations. The buildings may be different. The business processes may be different. Certainly, the people will be different. It is important to plan an architecture that meets the needs of each IoT deployment. The good news is, that most such changes are required in the design for an implementation of an architecture and not in the architecture itself;

---

[138] Adapted from Ameyed, Petrillo, Jaafar, et. al. *Internet of Things Architectures: A Comparative Study.* 2020

however, as an IoT professional, you should analyze the architecture for each deployment to ensure that it indeed meets the requirements.

The basic process for converting a reference architecture into a concrete architecture is as follows:

1. Evaluate the requirements documentation to determine the conceptual components that must exist in the network and note them.
2. Create a drawing of the network, application, and data storage/processing components showing their relationships with one another.
3. Define the interfaces for each component (APIs, network connections, database connections, etc.).
4. Define all concepts, services, components, technologies, and other items represented in the architecture drawing[139].
5. Perform architecture analysis (flow analysis to identify link requirements, QoS, reliability, etc.; risk analysis to identify failure points, security concerns, etc.; requirements analysis to ensure all requirements can be met with the architecture).
6. Communicate the architecture to the stakeholders, where appropriate, and gain approval from the acquirer.

The input to the architecture development process is the output of the requirements engineering process. Depending on the documentation created during requirements engineering, a potential architecture may already be described. For this reason, the first step in creating the concrete architecture is to evaluate the requirements. In some projects, you performed the work documented in chapters 3 and 4 and created the requirements yourself. In others, you are handed the requirements and asked to design the solution. In either case, start with the requirements. If the requirements are insufficient, clarify needs before moving forward.

With the conceptual components ready, you can begin to create the architecture drawing. Any number of tools can be used for this, but in this case, auto diagramming tools (those that scan the network and build a diagram) will not help. The network does not exist, so you must use a tool that allows you to manually create the architectural diagram. These

---

[139] An architecture is more than just a diagram. The document will (hopefully) include one or more diagrams, but it should also include definitions of terms, concepts, components, interfaces, and so on.

tools include flowchart-type tools, graphics design tools, and presentation tools. The following list should get you started:

- Diagrams.net (Draw.io, Free, can be run on a local server)
- Microsoft Visio (Commercial)
- eDrawMax (Commercial)
- LucidCharts (Free, up to 3 documents, the licensed monthly)
- ConceptDraw Diagram (Commercial)
- Network Notepad (Free limited version, Commercial version)
- LanFlow Net Diagrammer (Commercial)
- yEd Graph Editor (Free)
- Visual Paradigm (Commercial, Limited Free)

An architecture diagram does not have to look like a traditional network diagram with icons of servers, switches, routers, gateways, etc. It can use simple blocks with names indicating the concept or function performed. In fact, when creating the design from the architecture, it is entirely possible that a single server will perform several architectural functions. It is also possible that a single architectural function will be spread across multiple servers.

As an example, consider an architecture diagram that specifies that MQTT will be used as a message broker, or simply that a message broker will be used to receive information from IoT devices and send information to applications. In actual design and implementation, you may choose to implement MQTT-SN with eleven forwarders and two gateways spread across thirteen total devices. The architecture simply had a box that read "message broker service," but the design accounted for the distribution and number of end devices and implemented thirteen physical components to meet the architectural and system requirements.

With the components drawn, you can indicate the relationships between them in the drawing. In addition to the drawing, it is advised to document the interfaces in sufficient high-level detail to ensure they are designed properly. For example, you may choose to define that all API interfaces throughout the system shall be REST APIs. Such an architectural constraint guides the design process.

In addition to documenting the interfaces, all concepts, services, components, technologies, and any other entities in the diagram should be documented in text. The text will act as a support for the diagram to ensure that the designer understands the intention. You may be the architect and the designer but documenting it in this way also provides for simpler future support, scalability, and upgrades.

The last step before communication with the stakeholders and acquirer is architecture analysis. This step involves inspection of the architecture to ensure that it truly meets the requirements of the system or systems. Flow analysis involves mapping the flow of different types of communications through the architecture and ensuring that it will allow for required performance, reliability, security, and other factors defined in the requirements. Risk analysis involves inspecting the architecture for security concerns and potential failure points that would prevent the system from accomplishing quality requirement goals. Any identified risks should be removed or mitigated. Mitigation may be as simple as defining a plan for occurrence or it may involve implementing redundancies, additional security, etc. Finally, the last step is requirements analysis of the architecture. One last evaluation of the requirements and a final determination that the architecture, if implemented with a proper design, will fulfill all requirements.

The output of the architecture development process is a comprehensive architecture that can be used by the wireless IoT designer to plan and design any requirement components. This architecture should be reviewed with the acquirer and key stakeholders, when appropriate, and approval should be gained for moving to the design stage.

Keep in mind that the architecture will not typically document the "where" of the network, but it will document the "what" and some of the "how" of the network. The designer will take this architecture and the requirements and then design the "where" component as well as implement through design the "what" and the "how." Stated differently, the architecture documentation is not the source of information defining coverage areas and requirement areas on the wireless side. That information will come from the requirements.

In many cases, while the architecture typically remains at a high level, it will define the wireless IoT protocols used for access by end devices. Purists would be opposed to this as the architecture has traditionally been constrained to showing only the relationships and not the specific technologies used on the links; however, depending on the size of the network, the architecture may begin to blend with the design. As an example of traditional differences between architecture and design, consider Figure 5.20. Whether the understanding is needed in architecture or design development, it is important to understand the wireless IoT topologies and protocols when planning the architecture. The remaining sections of this chapter address these issues.

Architecture	Property	Design
Broad ←	Scope	→ Focused
Generalized ←	Detail	→ In-Depth
Relationships ←	Description	→ Technologies
Independent ←	Location	→ Dependent

Figure 5.20: Architecture vs. Design

5.4: Wireless IoT Topologies

The different IoT protocols support varying topologies and architectures. This section will provide a review of several common topologies so that you have them in mind when a specific protocol references them and as you design the wireless IoT solution.

Point-to-Point (PtP) and Peer-to-Peer (P2P) Topologies

A subtle difference exists between PtP and P2P in some IoT protocol descriptions. PtP indicates that the two devices have a link with each other that is protected such that other devices cannot view the conversation. P2P is different in that it allows communications that others could view as it is transmitted in the same network setup as all other communications, whether encrypted or not, like a broadcast, but sent to a specific device. For example, in the image for PtP, notice that two devices have a link that only they can view and two other devices, though in the same air space, have a link that only they can view. Whereas, on the P2P side, all devices can see all the links. They simply agree to only "receive" what is intended for them.

The concept can be even more confusing because the use of the term *peer* in many standards simply references any other device with which a given device is communicating using the same protocols. For example, Zigbee defines a network protocol data unit (PDU) as "the unit of data that is exchanged between the network layers of two peer entities." Therefore, it would be accurate to say that the two entities are in a peer-to-peer relationship. In the Bluetooth specification, it is stated that "in the BR/EDR core system, *peer devices* use a shared physical channel for communication." Again, this description would define the link as a peer-to-peer link. However, the Bluetooth specification also states that "Inquiring and discoverable devices use a simple exchange of packets to fulfill the inquiring function. The topology formed during this transaction is a simple and transient *point-to-point connection*." Therefore, ultimately, the specification defines a point-to-point, point-to-multipoint (piconet), and peer-to-peer topology, even if it doesn't declare it specifically[140].

Figure 1.12: PtP and P2P Topologies

Point-to-MultiPoint (PtMP) Topologies

A point-to-multipoint (PtMP) topology has a central device that communicates with several other devices. The central device can send broadcasts that reach all other devices. The central device (point) may also transmit to specific devices; however, unless some method is used to protect the communication for unicast purposes (such as peer-to-peer encryption), the other devices could view the communication even though it is not intended for them. When unicast encryption is used, all devices could demodulate the wireless communication, but they could not decrypt the payload and, therefore, could not view the useful information in the exchange.

A Bluetooth piconet with devices and a single master is an example of a point-to-multipoint topology. Traditionally, PtMP as a topology name has been used for mostly smaller networks. However, the star topology is ultimately a large-scale version of PtMP.

---

[140] The CWIDP exam does not test on the subtle differences between PtP and P2P as they are often used interchangeably within the industry.

Figure 1.13: PtMP Topology

Star Topology

The star topology, in data networking, has been referenced for several decades. It is challenging to define the difference between star topologies and PtMP topologies. Some suggest that the only reason these different topologies exist is because wireless was introduced and used the term point-to-point to reference two wireless nodes communicating only with each other and, therefore, one wireless node communicating with several others must be point-to-multipoint – instead of star.

Whether you differentiate them or not, the most important thing to note is that some IoT specifications use the term PtMP in the way that others use star. Therefore, for the purposes of the CWIDP exam, you can think of the two as roughly equivalent. The star topology uses a central "hub" to which all remote devices communicate.

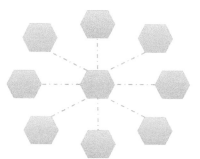

Figure 1.14: Star Topology

Star-of-Stars (Extended Star) Topology

Star-of-stars topologies implement a local star that communicates with one or more remote stars. For example, LoRaWAN is defined as a star-of-stars topology. The local

LoRa devices communicate with the LoRaWAN gateway (first star) and the gateway forwards the messages to a network server where multiple gateways can forward messages from their LoRa end-devices to the same network server (second star). Finally, the network server may be connected to another network with multiple application servers in communications with it (third star).

The LoRaWAN specification defines it like this: *LoRaWAN networks typically are laid out in a star-of-stars topology in which gateways relay messages between end-devices and a central Network Server the Network Server routes the packets from each device of the network to the associated Application Server.*

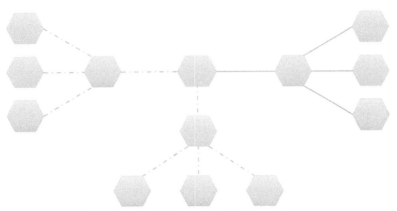

Figure 1.15: Star-of-Stars Topology

Tree/Cluster Tree Topology

The two popular IoT protocols that define a tree topology are 802.15.4, which references trees and cluster trees, and Zigbee, which references only trees. The Zigbee specification specifically states that it only defines intra-PAN networks; therefore, cluster tree networks are not defined, though the standard does not constrain a custom implementation that might utilize them. For example, Digi XBee is a proprietary implementation of Zigbee that could allow for a network resembling a cluster tree topology due to that fact that all nodes on the DigiMesh are equal peers with full capabilities for routing. Of course, Zigbee is itself based on 802.15.4.

The concept of the tree is simply a top-down or bottom-up hierarchy where all nodes eventually reach the root of the tree. The cluster tree expands this by allowing multiple tree topologies to interconnect.

In some standards, the term cluster tree also refers to a single tree.

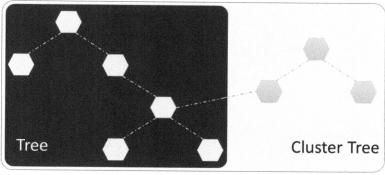

Figure 1.16: Tree and Cluster Tree Topology

Mesh Topology

A mesh topology allows for devices to establish connections with multiple other devices and for routing or forwarding to take place through the mesh. In a *full mesh*, all nodes connect to all other nodes. Obviously, a full mesh is only possible within specific constraints, for example all nodes must be able to establish a link with all other nodes. Alternatively, a partial mesh may be implemented that allows for all nodes to have some path available to them through the mesh and to other nodes in the mesh though all nodes are not connected to all other nodes.

Additionally, some nodes may participate in building the mesh itself, while other nodes simply connect to the mesh to transfer messages through it. Therefore, a mesh node, is one that participates in routing or forwarding within the mesh and an end-node or mesh client would be one that only uses the connection to the mesh through a mesh node to send messages. Some specifications refer to devices capable of building the mesh and participating in routing or forwarding as full function devices (FFDs) while those that can only act as end devices are reduced function devices (RFD).

To calculate the required links for a full mesh implementation, use this formula:

227

Number of Links = Nodes x (Nodes − 1) / 2

For 8 nodes, the formula is:

Number of Links = 8 x 7 / 2 = 28

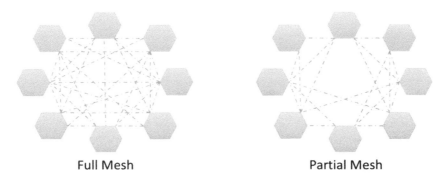

Full Mesh Partial Mesh

Figure 1.17: Mesh Topology

LPWAN Topologies

Low Power Wide Area Networks (LPWANs) form long distance links for IoT devices. They introduce a new phenomenon to the concept of a WAN in that traditional WAN links are long-distance links between networks while LPWAN links can be long distance links directly from end devices to the remote network in most IoT deployments. For example, with cellular IoT, a device can link directly to the cellular network with a tower location being several kilometers away. The same is true with Sigfox or LoRaWAN, where an end device may connect to the Sigfox network or a LoRaWAN gateway that is several kilometers away.

At the same time, an LPWAN topology may allow devices to connect to gateways in closer proximity and then the gateway connects to the long-distance network. Either way, the concept of the LPWAN is the ability to form long distance connections. A protocol is typically considered an LPWAN protocol if it can create links at a distance of one kilometer or longer.

Within the industry, some define cellular as an LPWAN protocol/technology, and some do not. Additionally, some refer to protocols like LoRaWAN and Sigfox as non-cellular

LPWAN protocols, which suggests that cellular is also an LPWAN protocol. For the purposes of this course and the CWIDP exam, cellular is also considered an LPWAN protocol.

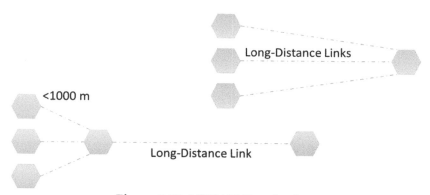

Figure 1.18: LPWAN Topologies

Long Range vs. Short Range Wireless IoT

An important distinction in IoT protocols is the range in which communications can occur. Some are long range protocols and others are short range protocols. Range, in free space, is primarily a function of frequency (f), channel bandwidth (bw), transmit power (p), antenna gain (both transmitter (gTx) and receiver (gRx)), receive sensitivity (s), modulation (m), and coding (e).

$$R = f(f, bw, p, gTx, gRx, s, m, e)$$

Lower frequencies with longer wavelengths can typically be received at greater distances. Wider channels are more difficult to receive at greater distances. Higher output power allows the signal to be received at a greater distance. Antenna gain helps to increase distance by focusing both transmit directionality and receive directionality. The components in the radio can result in improved or degraded receive sensitivity, which impacts the required signal strength (and, therefore, range) in communications. More complex modulation requires better signal strength (and, therefore, shorter range) than less complex modulation. Coding can provide more, or less, redundancy and resiliency of data and therefore increase or decrease range.

In the end, long range communications can sacrifice some of these parameters, but never all of them. For example, if low power is desired with long range links, narrower channels and less complex modulation are typically used with more redundant coding methods. In fact, using narrow channels and resilient modulation techniques is a specific characteristic of LoRaWAN and Sigfox that helps to provide their range. The longest range is achieved with the lowest data rate.

Choosing Connectivity Protocols

Many factors go into selecting the proper wireless protocols to connect your IoT system. Capacity is an important piece of the planning and selection mechanism. For unlicensed spectrum, the constraints that need to be thought about most are:

- Number of devices
- Messages per day per device
- Message Size
- Distance
- Data rates

Table 5.1 provides a high-level summary of the protocols covered in the CWNP IoT track. The CWIDP exam will not test your knowledge of specific ranges for IoT protocols, though you should know the bands used. With varying antennas and mounting heights, it is entirely possible to achieve longer ranges than those listed in the table. It is also possible to achieve shorter ranges.

This table represent data for the protocols mentioned and can be helpful when selecting the protocol to utilize for your IoT system. There may be additional factors that go into the decision-making process in environments with existing wireless IoT systems or additional concerns.

Protocol	Frequency	Range (Single-Hop)	Data Rate	Topology
802.11 (Wi-Fi)	2.4 GHz, 5 GHz, 6 GHz	30 – 2000 m	High	PtMP/Star
802.15.4	2.4 GHs/Sub-1 GHz	30 – 10,000 m	Low/Medium	PtP/Tree
6LoWPAN (Thread)	2.4 GHz	30 – 300 m	Low/Medium	Mesh/Tree
Bluetooth	2.4 GHz	10 – 1500 m	Medium	PtP/Mesh
ISA100.11a	2.4 GHz	30 – 300 m	Low	PtP/Mesh/Star
LoRaWAN	Sub-1 Ghz	100 – 15,000 m	Low	Star
Sigfox	Sub-1 GHz	100 – 15,000 m	Low	Star
WirelessHART	2.4 GHz	30 – 300 m	Low	PtP/Mesh/Star
Zigbee	2.4 GHz/Sub-1 GHz	10 – 300 m	Low	Tree/Mesh
Z-Wave	Sub-1 GHz	10 – 300 m	Low	PtP/Star

Table 5.1: IoT Protocols Covered on the CWNP IoT Track

5.5: Chapter Summary

In this chapter, you learned the details of architecture development. You explored a reference architecture framework in detail and learned basic information about two additional frameworks or models. Then, you learned how to create a concrete architecture. Finally, you learned about the common network topologies and the wireless IoT protocols that use them. In the next chapter, you'll begin designing the wireless IoT solution, with a focus on designing the wireless network. In later chapters, you'll explore the planning for other portions of the IoT architecture implementation.

5.6: Review Questions

1. How is a reference architecture framework used in relation to an IoT solution?

    a. It acts as the architecture for implementation

    b. It is used as a model to build a concrete architecture

    c. It is not used in IoT networks

    d. It is used to provide the minimum that must always be implemented

2. What are the three layers of the three-layer IoT architecture model?

    a. Sensing, Actuating, and Controlling

    b. Sensing, Actuating, and Communicating

    c. Perception, Network, and Applications

    d. Physical, Data Link, and Transport

3. What is the goal of the IoT-A IoT Domain Model?

    a. To provide a common understanding of the IoT world.

    b. To provide a concrete architecture for Windows Active Directory domains.

    c. To provide a concrete architecture for Internet domains

    d. To provide a concrete architecture for DNS services

4. True or False: The IoT-A Information Model models the information related to a Virtual Entity.

    a. True

    b. False

5. The IIC develops a Reference Architecture for the Oil & Gas industry that is not applicable to other industries.

    a. True

    b. False

6. What phrase is used for the architecture of a specific system that may or may not be based on an existing model?

    a. Reference Architecture Framework

    b. Concrete Architecture

    c. Architecture Model

    d. None of these

7. When creating an architecture for a specific solution, what should be performed immediately before stakeholder review and approval?

    a. Architecture analysis

    b. Requirements engineering

    c. Project budget establishment

    d. None of these

8. What is a common maximum range of long-range wireless IoT protocols?

    a. 10 meters

    b. 100 meters

    c. 1000 meters

    d. 10,000 meters or more

9. True or False: A full mesh topology has interconnections between every node.

    a. True

    b. False

10. What kind of wireless IoT protocols is designed for long-range connections?

    a. Zigbee

    b. LPWAN

    c. WirelessHART

    d. ISA100.11a

5.7: Review Answers

1. The correct answer is B. The reference architecture is used as a model to build a concrete architecture.

2. The correct answer is C. The three-layer model uses perception, network, and applications.

3. The correct answer is A. The Domain Model, models the IoT domain of concepts and knowledge and therefore provides a common understanding of that world.

4. The correct answer is A. The statement is true.

5. The correct answer is B. The statement is false.

6. The correct answer is B. The statement defines a concrete architecture.

7. The correct answer is A. Architecture analysis can reveal problems in the architecture or missing components.

8. The correct answer is D. The common range is 10 kilometers or 10,000 meters or more.

9. The correct answer is A. The statement is true.

10. The correct answer is B. Low Power Wide Area Network (LPWAN) protocols are designed for long-range connections.

Chapter 6: Designing the IoT Wireless Network

Objectives Covered:

3.2 Design for appropriate channel configuration

3.3 Design based on RF requirements and capabilities

3.4 Use wireless IoT tools to create and validate the design

3.5.3 Recommend robust security solutions

The design[141] process is the part of the IoT solution project that ensures all the concepts, components, devices, protocols, data, and any other pieces of the IoT puzzle work together to achieve the requirements of the solution. While it is interesting, maybe even exciting, to see wireless technologies work, in the business world, if the system doesn't meet the needs of the organizations, it's a failure. Figure 6.1 illustrates the importance of getting the pieces right. Here, as you can see, the puzzle pieces do not line up to form a fully functioning system. The protocol does not provide the capabilities required of the system and the puzzle piece doesn't fit. While there are dozens of wireless IoT protocols, we represent only three in the image. They can be any three you like, but the point is simple: If the protocol does not meet the latency, resiliency, security, throughput, range, and other requirements of the system, the other components can be perfectly designed, and the system will still fail.

 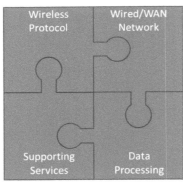 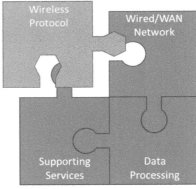

Proper Design   Improper Design

Figure 6.1: Getting the Right Puzzle Pieces

---

[141] The *American Heritage Dictionary*, 1991, defines *design* as 1) To conceive in the mind; invent. 2) To form a plan for. 3) To have a goal or purpose; intend. 4) To plan by making a preliminary sketch, outline, or drawing. 5) To create or execute in an artistic or highly skilled manner. All these definitions have applications to our wireless IoT designs. To conceive something in the mind, one must understand concepts related to that thing. Once conceived, it can be planned with a goal or purpose in mind. This plan can be taken out of the mind and documented with lists, drawings, and other artistic representations. Yes, wireless IoT design is a science (knowledge and skills are required), but it is also an art (creativity and inventiveness are required as well). Interestingly, the noun definition provided in the same source, *a decorative or artistic work*, even applies to our task as aesthetics become an important part of many projects.

This chapter is focused on the design of the wireless IoT network. It is about getting that puzzle piece right. The next chapter will focus on the other puzzle pieces, though not with the same depth of coverage given to the wireless network, given that it is our primary priority in this text. For example, while this chapter is wholly focused on the wireless network design, the next chapter will cover planning for the wired network, the end devices, supporting services, IoT Data protocols, data storage and streaming, application planning, and monitoring and administration. Clearly, we are placing a priority on the wireless network design.

6.1: A Design Framework

Let's begin with a basic framework for the design process. This framework assumes you have proper requirements that are documented, and a concrete architecture defined. Therefore, it begins with the inputs of system requirements and architecture and ends with a functional design for the wireless IoT solution. Do not think of the framework as a sequence of events from left to right, but rather as the major tasks that must be performed to successfully design an IoT solution. Figure 6.2 illustrates the design process framework we recommend.

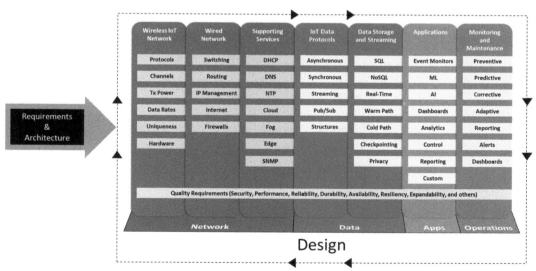

Figure 6.2: The Design Process Framework

Like any other framework, you may select from it the components required for a given project and add to it that which is lacking. It simply provides a framework for the most common design tasks and concepts for consideration within the process. The tasks of wireless IoT solution design include[142]:

- Wireless IoT Network Design
- Wired Network Design/Planning
- Supporting Services Design/Planning
- IoT Data Protocols Design/Planning
- Data Storage and Streaming Design/Planning
- Applications Design/Planning
- Monitoring and Maintenance Design/Planning

These tasks are collected into four groups: Network, Data, Apps, and Operations. For the CWIDP, the Network group is the most important. However, the remaining three groups cannot be ignored as they impact decisions made within the Network group. For example, if the Operations team wants to capture information using SNMP traps for monitoring the IoT gateways, these configurations must be implemented within the wireless and wired networks through the SNMP supporting service.

Within each task is a list of concepts or technologies that must be considered for design and planning. Additional concepts or technologies may be required in specific projects; however, those listed are sufficiently common to warrant inclusion. As you design for each task or task group, ensure that the listed components are addressed, and any additional required components are addressed as well. Again, this is a framework, which means that your actual project may include more or fewer items than those listed in Figure 6.2.

At first, it may appear that some items are missing from the lists. Consider the example of frequency. A quick glance at Figure 6.2 reveals that we have not included the term frequency or the phrase frequency band in the list. However, notice that *protocols* and *channels* are in the list. Indeed, the protocols you select to use and the channels on which

---

[142] Note the use of the word *planning* for all but the wireless IoT network design. This change is intended to communicate the level of coverage we provide in these materials. More focus is given to the wireless side of the design than the others. This focus is not intended to diminish the importance of the others, but full coverage of them all is beyond the scope of a single book or certification.

the network will operate determine the frequencies or frequency bands that will be implemented. It is assumed that this framework will be used by a professional who understands regulatory constraints for the target region. Therefore, regulatory agencies and regulatory control are not directly listed either.

Another example is the Hardware component in the Wireless IoT Network task. This is inclusive of end devices, gateways, FFDs, RFDs, and basically any device that is part of the wireless IoT network. Generally speaking, to design the wireless IoT network, you will start with the end devices, understanding what is required by them, and then implement the wireless IoT network design to meet these requirements. You do not typically start by choosing a wireless IoT protocol, like ISA100.11a, and then ask, "How do I make these devices work with this?" Exceptions do exist, of course, such as when an existing wireless IoT network is in place and the organization wants to add end devices to that network. The goal was to provide a framework that motivates the consideration of the important factors, which will, in turn, lead to other factors of which the professional is aware.

One final note about the framework should be made. Notice that one component is in every design task within every group: Quality Requirements. Security should be a consideration in every element of the design. Using this approach results in better security overall. As stated previously within this book, security is a cross-cutting or horizontal requirement. If you consider the wireless IoT network, wired network, supporting services, IoT Data protocols, etc., as vertical design tasks, security and other quality requirements drive horizontal design tasks. This is not unusual to a quality requirement imposed on the design. To achieve these quality requirements, they must be considered within each component of the system because the goal is end-to-end security, end-to-end QoS, end-to-end reliability, etc.

The remainder of this chapter will explore the components within the Network group of tasks including the wireless IoT network, wired network, and supporting services. First, we must address the issue of "build or buy" as we discuss selecting and building IoT solutions.

6.2: Selecting and Building IoT Solutions

Before investigating the specifics of each design task in the Network group, the topic of "build or buy" must be addressed. Deciding if you should buy an off-the-shelf IoT solution or if you should build your own is sometimes a difficult task. In other situations, it is easy. The first step is to understand what solutions are available within the target set of IoT solutions for the specific industry in which you are deploying. After the solution providers have been identified, you can determine the features these solutions provide and link them with the requirements and identify any potential gaps in meeting those requirements. If there are gaps, the stakeholders may opt to forego a requirement to have a solution that they can deploy quickly, or you may have to build an entire solution or some components to fill the gaps.

If the solution doesn't exist today within the ecosystem partners, building a custom IoT solution may be the only way to meet your requirements. In this case, the budget needs to be assessed to determine if this is feasible and compared with the available talent and skillsets required to develop such a product. Off-the-shelf hardware and software may be available, or hardware and software may need to be custom built to meet the requirements. Creating a budget and timeline can be difficult, with many unknowns in the early stages of a development project. If both hardware and software need to be created some tasks will need to be done in serial as you cannot fully test and validate software functions on hardware that doesn't exist. An entire book can be written on building solutions from the ground up; however, that is not the purpose of this certification.

The requirements will lead you to a buy or build decision in most cases, along with a conversation with stakeholders. In the case of our panic button for hotels, there are several solutions available on the market and these companies focus on just this solution. TraknProtect and React Mobile are two such providers with deployments that meet a set of requirements. With your list of requirements, you can compare them against what these companies provide and decide whether to build or buy.

Earlier in the requirements gathering we discussed having a map or site plans with location requirements for coverage. You can use these plans to lay out the IoT network connections. Once the network is designed, a proof of concept can be performed to validate the system meets the requirements. Not all systems will require a proof concept if they have a proven solution and track record of delivering results.

You should now understand the costs to deploy the off-the-shelf system compared with a custom-built solution. You can compare these costs and the ability to meet requirements to decide what is best for the IoT system being deployed.

6.3: Wireless IoT Network Design

Designing the wireless side of the IoT network and solution will demand an understanding of the end devices required. The end devices are always the starting point as they are the reason the network exists. After the end devices have been selected, the supported gateways can be selected. These first two decisions, the end devices and gateways, typical make the third decision for you: the protocols. Next, you will determine the channels and transmit power to use and the data rates that can be achieved. Additional unique characteristics of the chosen protocol may also impact the design. All this information helps you to define the number of gateways required to reach the needed capacity. Finally, you must ensure the protocol is implemented such that it achieves the quality requirement of the project.

End Devices (Hardware)

Selecting end devices begins with the need. Do you need a sensor to monitor a physical entity or an actuator to control it? Are you simply using environmental sensors to monitor the environment? Do you require motion tracking devices to determine activity in an area? Whatever the need, this drives the sensor or actuator selection. Next, you must define the IoT characteristics of that device. Sensors have been around for decades as have actuators. The difference now is the connection of these devices to IoT networks instead of industrial control networks or simply not connecting them at all.

After the sensor, actuator, or tag need is identified, which should come from the requirements documents, you will have to select one that can participate in the final IoT network. Selecting IoT end devices includes considering the connectivity model, the communication protocols, power options, security, and machine intelligence, as depicted in Figure 6.3[143].

---

[143] Of course, other factors may apply in specific scenarios. As a more general guide to wireless IoT network design, we must constrain the topics to those most commonly required.

- **Connectivity Model:** At a high-level, will the IoT devices connect with a wired or wireless connection. Just because you are designing a primarily wireless IoT solution does not mean that all IoT end devices should connect via wireless. Determine those cases where wired connections should be used and specify them in the design.

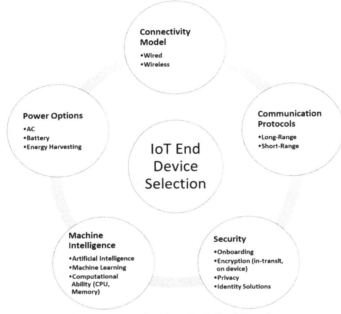

Figure 6.3: Selecting IoT End Devices

- **Communication Protocols:** Will you require long-range protocols, like LoRaWAN, Sigfox, LTE/5G, or will short-range protocols suffice, like 802.15.4, Bluetooth, or Z-Wave.
- **Power Options:** How will the devices be powered? Most of the devices using the wired connectivity model will likely be powered either with AC or PoE (for example, the gateways and end devices not demanding a wireless connection). Many devices may require battery power as they are mobile or too far from an AC power source. Some devices, requiring very long battery life while using protocols that consume more energy, may also need energy harvesting solutions

to be employed. Additionally, devices may be attached to other equipment that can provide the power such as vehicles and other machinery.
- **Security:** How will the devices be onboarded, what encryption is supported for transmission and any on device storage, how will privacy be protected (it's more than encryption, it's also about not sending unneeded information), and how will authentication occur (identity solutions)?
- **Machine Intelligence:** If on device AI or ML is required, this will significantly impact the compute requirements on the end devices as well as memory requirements. It will also circle back and have an impact on power options. If compute can be performed in the gateways, applications, or cloud, this is preferred as it reduces the cost of end devices and device management; however, if it must be performed at the end device, the level of compute should be minimized so that only what is required is performed.

With these parameters identified, the search for the right end device (either purchased or built) can begin. Figure 6.4 illustrates the basic components of an IoT end device. We will explore the important components that are related to the selection criteria previously explained.

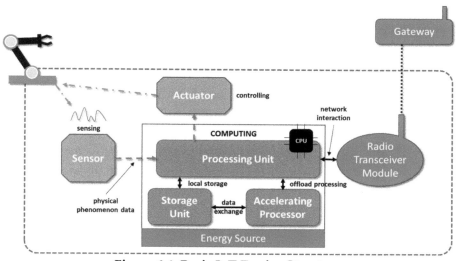

Figure 6.4: Basic IoT Device Structure

The first component to note is the *energy source*. IoT devices can be powered by battery, energy harvesting, battery and energy harvesting, or AC (mains) power[144]. The energy source should be planned to include the following:

- **Type of energy source:** battery, battery plus energy harvesting, energy harvesting, AC (mains).
- **Battery selection:** only quality batteries with long life should be used. The long life is in reference to both the available energy and the durability of the batteries to avoid other breakdowns.
- **Backup energy source:** for mains powered solutions, batteries are often provided as a backup in case the AC power goes down.

The *processing unit* is the main CPU in the IoT device. Many IoT devices can use very low frequency CPUs (measured in megahertz) and do not require single- or multi-gigahertz processors. However, on a case-by-case basis, the required processing power must be determined. If the device will run local compute processes more intense than simply reading sensor data or sending commands to a physical entity (such as a motor or valve), a more powerful CPU may be required. This latter requirement is mostly the case when on device decision-making is required or lightweight AI/ML algorithms are executed.

Some devices will also have an *accelerating processor*. This processor is dedicated to tasks such as video processing or signal processing that requires intensive algorithms. The accelerating processor is also called and offload processor as certain types of calculations, for which it is best, are performed there instead of on the main CPU.

A *storage unit* is required on all end devices. In most cases, it is simply a flash ROM type unit used to store the software that operates the device. However, with some IoT devices, such as video cameras, local storage may be used to store the video for local processing with an accelerating processor before transmission in a file transfer format rather than a stream format. In most cases, streaming transfers are used on such devices. If any other on device processing is performed, the storage unit may require more capacity as well so that the CPU or accelerating processor may perform computation against the data. This latter method can be quite beneficial when the device harvests a large amount of data but

---

[144] Some devices are powered through a form of on-demand energy harvesting where the push of a lever or button generates significant power to radiate a brief message.

need only transmit that which is essential after local processing; however, it will also require more energy.

Attached *sensors* transmit *physical phenomenon data* to the device (or it will be read from the sensor, depending on perspective and operations), typically in the form of electrical signals. The device will translate these signals into meaningful data and send them across the network for some intended target. The most important factor here is selecting devices with sensors or selecting sensors for devices that sense the required phenomena and sense it in the range required. For example, if you must detect temperatures up to 100 degrees Celsius, a sensor that only goes up to 90 degrees will not work.

Attached *actuators* will receive controlling commands from the device to turn something on or off, to change a flow rate, to move a robot arm, etc. The device must be able to translate commands from the network into commands the actuator can process. Alternatively, if all commands are generated on the local device based on sensed information on the same device or another device communicating with it, the device may simply generate the controlling commands for the actuator locally.

The final component in the end device is the *radio transceiver module(s)*[145]. This module is responsible for radio communications and must support the protocol desired. Some devices will incorporate multiple modules, for example, an 802.15.4 radio and a cellular radio. The device may be configured to use one or both modules in a backup/fallback configuration.

Constrained-Node Network (CNN) and RFC 7228
The IETF published an informational RFC in 2014 that defines terminology for Constrained-Node Networks (CNN) that consist of constrained nodes or devices. These devices are defined as *small devices with limited CPU, memory, and power resources*[146]. The document classifies these constrained nodes into three classes:

- Class 0, C0: RAM less than 10 KiB and code size (flash) less than 100 KiB
- Class 1, C1: RAM approximately 10 KiB and code size approximately 100 KiB

---

[145] In addition to the information discussed here, the antenna should be considered as well. Some devices have no option for an external antenna and others do. If they do provide an external antenna connection, using different antennas can allow the end device to reach the gateway or the mesh network without requiring that additional routers, mesh nodes, or gateways be added to the network.
[146] IETF RFC 7288 Terminology for Constrained-Node Networks

- Class 2, C2: RAM approximately 50 KiB and code size approximately 250 KiB

They further define a C0 device as very constrained sensor-like motes. They typically cannot communicate directly with the Internet because they lack the memory to implement the required protocol stack. Proxies, gateways, or servers are required to allow communications beyond the simple, local wireless link.

A C1 device is also constrained but may be able to implement communications with Internet technologies such as CoAP with UDP due to the limited stack requirements. However, it would still lack the capacity to implement TLS, HTTP and other security protocols or complex data representations.

A C2 device is the least constrained and, believe it or not, likely has the ability to implement a full stack for network communications much like a laptop computer. However, it is still constrained as to the size of RAM and of code storage and so it will be hindered from performing on device AI, ML, and other advanced processing. For such tasks, an unconstrained IoT end device would be required. It may still have a processor running at less than 100 or 200 MHz and it may only have a few megabytes or RAM with storage for several megabytes of code, but when a device is dedicated to a purpose and a large, bloated operating system is not required, much can be accomplished with a little. Such devices may still be constrained, but RFC 7228 does not address them specifically.

End Device Use Cases and Applications

It is important to understand some of the more advanced use cases and applications related to the end devices as they may require further planning. These applications include:

- Digital Twins
- Robotics Applications
- UAV Applications

Digital twins are often used to mirror physical things in the digital or virtual world. A digital twin differs from a simulation in that a simulation exists in a completely virtual world with no physical counterpart. A digital twin is linked/connected with a physical counterpart. As an example, consider using design software and placing gateways on a floorplan. You can configure the transmit power and antenna type as well as the channel and then simulate the results in the virtual world to model the physical world. Now, imagine gateways are deployed in a real environment and both the gateways and the environment are modeled in the management system. You can now change settings on the digital twin in the management system, simulate the results and, if acceptable, make them real in the physical world. We have not called it a digital twin in many systems, but the concept is there, even if it is most often more of a digital shadow, which is explained later.

The digital twin can model a physical object in software as a complete model, a partial model, or an enhanced model. While the first two are self-explanatory, the enhanced model requires further explanation. The enhanced digital twin is one that can do more than the physical counterpart, but the extra capabilities are possible given modifications to the physical counterpart. It allows you to test results should the modifications be made. Figure 6.5 shows the basic concept of the digital twin with the physical robot arm represented on the left and the digital twin represented on the right.

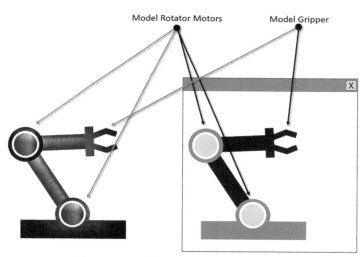

Figure 6.5: Digital Twins Illustrated

The benefits of using digital twin technology include change testing, interacting with real objects through the virtual version, and monitoring real objects as virtual entities. For example, if you look at the digital twin and it reports a temperature of 73 degrees Celsius, this is because the physical entity as at that temperature[147].

The phrase *digital shadow* typically revers to a uni-directional twin. In this case, either the physical entity reports status updates to the twin, or the twin controls the physical entity, but both are not performed – it is not bi-directional. For example, if the twin can send commands to turn something on or off, but not read the state of that thing from the physical entity, it is a digital shadow.

Robotics is an area of automation and potentially IoT. The robots may represent another common end device use case for wireless IoT. In this case, local computer power is much higher as many decisions are made at the device. Robotics is used in warehousing, industrial manufacturing, military, and even in healthcare applications, such as assistive robots for in-home care. Important considerations for these end devices include network interfaces (often wired for stationary robotics and wireless for mobile robots), data generation and reception (can be much higher than most other IoT devices), mobility capabilities, and application requirements such as ML and AI. The Robotics Operating System (ROS) is used in many applications to abstract common functions in robotics and allow developers to focus on getting things done rather than getting things to work. It was first released about fifteen years ago and is currently on ROS version 2. When you understand how robotics has evolved, you will quickly grasp that the autonomous vehicle is simply a special category of robotics, and we will be seeing more and more of them on the roadways in the next decade.

The final advanced use case/application we will discuss is *unmanned aerial vehicle (UAV)* applications. Figure 6.6 presents some of the common use cases. Additional use cases not listed are military and retail delivery. UAVs are highly mobile and may require cellular connectivity for full operation in a metro area or similar large region. They often include auto-navigation functions and IoT devices attached to the UAVs are powered from the batteries used to power the UAV. These batteries are recharged when the UAV returns to the base station. One of the most interesting use cases is that of the roaming gateway. The UAV takes the gateway to the end devices at time intervals to allow the devices to communicate with the network. The UAV either stores the data received from the end

---

[147] The concept of the digital twin varies significantly across research, implementations, and vendors. In many cases, it's nothing more than a JSON configuration set for the device that can be pushed to the device and configuration parameters can be read from the device.

devices until it returns to the base stations or it uses a longer-range wireless link to immediately forward them to the network.

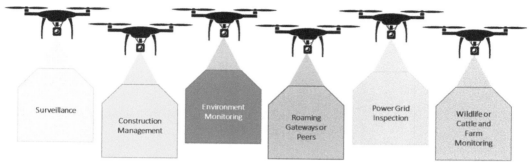

Figure 6.6: Common UAV IoT Applications

Gateways (Hardware)

We are using the term *gateways* here in the most inclusive sense as a reference to all devices that provide a gateway function. For example, the border routers of a 6LoWPAN network fit into the category and the actual gateways of a LoRaWAN network fit into it as well. The generic term gateway, in this case, is the device or devices that connect the wireless IoT network with other networks[148]. Gateways can be considered from at least three perspectives:

- The device to which wireless end devices connect.
- The device or service that connects to the Internet.
- The device or service that connects to a cloud provider.

All three of these scenarios apply in wireless IoT networks and other scenarios may apply as well. A conceptual gateway is represented in Figure 6.7. For such a gateway to function in each implementation, one input/output interface must support the protocol(s) of the local network and another input/output interface must support the protocol(s) of

---

[148] The Bluetooth Gateway Study Guide, published by the Bluetooth SIG, states that *definitions of the term "gateway" vary, largely according to the context in which the term is used. In general, though, a gateway is a system which allows two otherwise incompatible communications systems to talk to each other via a network or networks.* We completely agree that context drives the meaning of the term.

the remote network or the network through which the remote network is reached in a multi-gateway implementation.

Figure 6.7: Conceptual Gateway

Let's discuss each of the three common gateway perspectives referenced above. The first is the device to which wireless end devices connect. This perspective of a gateway is fulfilled by differently named components within different wireless IoT protocols. Table 6.1 presents the terminology used by various standards or protocols for the gateway component. As you can see, many use the term gateway and others use terms like coordinator, base station, border router, or more fully internet gateway (the Bluetooth SIG uses lowercase *internet* throughout their documentation for Bluetooth gateways).

Protocol	Term for Gateway "Type" Device
Bluetooth	Gateway/Internet Gateway
ISA100.11a	Backbone Router (Uses a gateway or is a gateway)
WirelessHART	Gateway
Zigbee	Coordinator[149]
LoRaWAN	Gateway
Sigfox	Base Station
6LoWPAN/Thread	Border Router

Table 6.1: Gateway Concepts in Different Wireless IoT Protocols

---

[149] Technically, the coordinator manages the Zigbee network only and a gateway is used to allow the centralized device to communicate outside the Zigbee network as well; however, in most cases, within the same devices that is the coordinator is the gateway. Therefore, from the perspective of the Zigbee end devices, the coordinator is the gateway.

When designing the gateway plan, multiple motives may drive the use of more than one gateway (or coordinator, base station, border router, or whatever the wireless protocol calls it). The following motivations are common:

- **Number of End Devices:** You have hundreds, or thousands, of devices and a single gateway will not perform sufficiently well with the large number of devices or the airtime will not be most effectively utilized with a single gateway.
- **Distribution of End Devices:** Whatever number of end devices you have, they are distributed over an area or areas too large to be serviced by a single gateway.
- **Backhaul Requirements:** Maybe a single gateway can both cover the area and handle the number of devices, but you want to prioritize the transmissions from certain devices over others. If the gateway or wireless protocols do not provide for this, it can be achieved by implementing dedicated gateways for the higher priority devices and configuring the network to grant higher priority to the traffic coming from that gateway (through QoS, traffic shaping, or some other methodology) than the traffic coming from the other. Of course, more than two gateways could be implemented for such a plan and more than two levels of prioritization could be used for the multiple gateways' traffic.

The device or service that connects to the Internet is also called a gateway, an Internet gateway, or an Internet router. All three terms reference the concept of providing connectivity from an otherwise private network to the public Internet. Such a gateway is required in wireless IoT solutions that connect to the cloud. It is referenced as a service because in some architectures the Internet connectivity is provided through a cellular provider and the actual connection from the cellular network to the Internet is unknown to the subscriber. It is simply referenced as a gateway service that the cellular provider makes available to the subscriber.

Finally, specific gateways may be used for application layer or IoT Data protocol communications to allow connectivity with cloud service providers. For example, gateways can be used to connect IoT edge devices to the Microsoft Azure, Google Cloud, or AWS cloud service providers' IoT service offerings. This allows the data to be forwarded to the cloud, processed in the cloud, and reported against. When designing an IoT solution that will be cloud-integrated, it is important to consider how the data will reach the cloud. Most providers offer two or three options:

- HTTP-based communications with no special gateway. Useful for constrained devices that need direct connectivity to the cloud.
- MQTT services with or without a special gateway. Allows MQTT-ready devices to communicate with the cloud.
- AMQP services with or without a special gateway. Allows AMQP-ready devices to communicate with the cloud.

Additional services may be supported by the selected cloud provider. The designer should verify with the cloud provider any options available. With proprietary protocols, such as Alta, used in Monnit IoT devices, the devices simply communicate with the iMonnit cloud directly if they are connected through a network with Internet connectivity.

Protocols and Localization

The hardware to be used (end devices and gateways) will determine the protocols to be used or the protocols to be used will constrain the hardware options. Indeed, it can work either way and so we presented hardware first, but you may select hardware based on a protocol choice or the other way around. Either way, when selecting hardware and protocols it will be based on the requirements and constraints of the network. Both the hardware and protocols must comply with these elements. With that said, let's discuss protocol selection issues, they include:

- Supported channels/frequency bands
- Transmit power requirements
- Data rates available
- Uniqueness (features unique to a protocol or not in all protocols)
- Localization requirements

The last one is important if it is required for your devices. We will discuss the others in the ensuing sections related to specific protocols. *Localization* is about locating the devices in physical space. Technically, any protocol can perform some level of accuracy for localization if the gateways will report to an application the signal metric information. At a minimum, they can be used to determine that a device either is or is not in communication with the gateway and, therefore, has moved out of the range of the network or is no longer communicating. However, when localization is required, this is rarely the kind of localization desired.

Figure 6.8: Radio Goniometer used at Ford Airport in the 1920s

Wireless localization has its roots in a term used less today, except for scientific circles, called *radiogoniometry*[150]. Radiogoniometry is the science of measuring radio wave reception source directions. The radio goniometer used at Ford Airport in the 1920s (Figure 6.8) was used in conjunction with receivers in the airplanes. It transmitted an "A" in one direction and an "N" in another. On the planes a receiver, could determine relatively precise location based on the received signals and fly toward the merging of the two.

---

[150] Pau, Arena, Gebremariam, You. *Bluetooth 5.1: An Analysis of Direction Finding Capability for High-Precision Location Services.* 2021. Sensors; Manna, Pelosi, Righini, Scorrano. *Compact Shaped Antennas for Wide-Band Radiogoniometry.* 2018. Applied Computational Electromagnetics Society Newsletter; Watt, Herd. *An Instantaneous Direct-Reading Radiogoniometer.* 1926. Journal of the Institution of Electrical Engineers; Poirer-Quinot. *Design of a Radio Direction Finder for Search and Rescue Operations.* 2015; Baltazart, Bertel, Fleury. *The Influence of Propagation on High Frequency Radiogoniometry and Single Station Location.* 1992. AGARD; Tatkeu, Berbineau, Heddebaut. *A New Approach to improve mobile localization based on angular deviation measurements in urban areas.* 1997. Conference Paper.

If you know the transmit power of a radiator and you know the antenna type used, then with a spinning goniometer radio direction finder (SGDF), you can locate the direction source of the signal and estimate the distance from the SGDF. In this case, using the SGDF as a receiver. To understand these early solutions to direction-finding, consider the following from the 1935 book, Measurements on Radio Waves and Antennas, Frederick Emmons Terman:

> *The direction in which a wave is traveling can be determined by the use of a receiver in conjunction with a loop antenna. The procedure is to rotate the loop until zero signal is obtained, for which condition the plane of the loop is perpendicular to the direction of wave travel.*

The position of the minimum rather than the maximum response is used since it permits much more accurate settings. Think of it like you turning your head to look in different directions. You are focusing the "antenna" of your eyes to pick up different electromagnetic waves (light waves). The direction that your eyes are facing, you intuitively know (as you're not really aware of "how" you know this), is the direction of the object seen. A house of mirrors reveals the limits of our radio goniometer (the eyes coupled with the brain).

These early direction-finding systems were known as "zero" systems of directive-reception and included the Bellini-Tosi System (using fixed aerials), the single frame system (rotating aerials), and the Robinson crossed-coil system (a multi-coil system). The phrase zero system was related to the fact that they were seeking a position of the antenna elements where the phase differences resulted in a null signal or zero signal. At that point, given half-wavelength separation, the line through the two antennas pointed in the possible direction of the transmitter (one of two directions). Measurements at other rotation locations (two for bi-angulation or three for tri-angulation) allowed for more precise determination of which direction. The different systems were spaced a few meters apart and could reveal the direction. Some attempts were also made at using RF blockers behind the antenna elements. Eventually better directional antennas could be used to determine general direction with a single radio goniometer.

Of course, all of this is sensitive to the environment. Early systems had problems at night that they did not have in the day. They worked well on open seas, but not in cities with many reflective objects. Additionally, the early systems didn't really have a method to measure signal strength as accurately as we do today, though signal strength measurements were first included in direction-finding as early as 1904.

Today, the word radar is so commonly used that we forget its origin in radio detection and ranging. It is an echo-ranging system wherein the radar transmitter sends out RF energy in a short burst of high energy and then listens for a reflection. The time delay of the returned echo is a factor of the distance traveled to the reflecting object and back again.

These early technologies are explained to assist in the understanding of localization solutions available in modern wireless IoT protocols. Localization is available in BLE, Wi-Fi, and several other protocols. If a device can be built with multiple antenna receivers or can collaborate with other receivers on the network, any protocol can theoretically support localization should higher layer applications be developed to process the received signal strengths from one or more receivers of the same signal, and if the location of those receivers is known as to their relation to one another.

In the following subsections, we will review BLE localization, Wi-Fi localization, RFID localization, and additional techniques. First, let's discuss the key performance indicators (KPIs) that must be met to achieve localization.

Localization for IoT

Localization is known as direction-finding, location tracking, real-time location/locating system/service (RTLS), and other terms depending on the protocol and technologies used. It is a collection of methods that determine the location of devices either for the network applications or for the device. For network applications, it is used to track the location of the devices. For the devices, it is used so that local applications can determine the device's location in relation to a facility or outdoor environment. It is used with IoT in healthcare for patient or equipment location tracking, in manufacturing for tool and product tracking, in hospitality for worker tracking and safety, and in many other application scenarios. Some applications require high levels of accuracy down to the centimeter and others work fine with accuracy levels measured in several meters.

The most familiar localization method used by millions is GPS in mobile phones and dedicated GPS devices. Based on GPS satellites, these localization systems are used for navigation and high-value location tracking. Other than navigation, GPS is only cost-beneficial for high-value location tracking because it is sometimes more expensive per location device (the costs include the tracker and the service that performs the location mapping) than other methods, such as BLE, Wi-Fi, RFID, etc. (the costs include the tracker (radio) and may use one-off purchased tracking appliances, though ongoing service subscriptions are required by some systems as well).

Localization is sometimes the primary driver of an IoT project. That is, the project is focused on location tracking as a solution and other IoT data collection is not a concern at the time of planning and deployment. However, when planning such a project, it is always a good idea to consider possible future projects and, when appropriate, plan a localization solution that could potentially be expanded for use with other IoT sensors and actuators, if required.

To achieve localization in wireless IoT solutions, several KPIs must typically be met. Some are required for functionality and others are required for quality.

- Network Coverage – Coverage is associated with the sensing range of the network. It can be considered the boundary within which a target may be located. Additionally, proper coverage by multiple "readers" may be required to provide accurate locationing. (Functionality)
- Security – Proper security helps to ensure availability of the localization service and accuracy in the reported information. (Quality)
- Target Recovery – When a targeted component is lost, due to coverage holes or sensor node failure, the system should provide rapid recovery and target relocation. (Functionality/Quality)
- Target Prediction – The ability to predict the location of a component based on probabilities. (Functionality)
- Localization Accuracy – The level of accuracy in localization from centimeters to meters. (Functionality/Quality)
- Energy Efficiency – The ability to perform localization with the lowest possible power consumption. (Quality)

Beyond the Exam: What is a Pose?

In robotics, localization is only part of the problem. In addition to determining the location, the robot's orientation must be determined. The combination of location and orientation is known as the *pose*. That is, the robot is at a particular location and is "facing" or "looking" or "moving" in a particular direction. In 2D space, the robot pose is often defined as an x,y axis with a degree measurement (offset from the parallel of the x axis on the 2D map), known as theta (θ). The "map" is known as a frame or coordinate frame. The horizontal scale (x) and vertical scale (y) map the locale of the robot. The entire pose is represented as a tuple of x_r, y_r, $θ_r$, where r is the robot under investigation. Therefore, thirty robots could have pose tuples with x_{r1}, x_{r2}, ..., x_{r30}, and y_{r1}, y_{r2}, ..., y_{r30}, and $θ_{r1}$, $θ_{r2}$, ..., $θ_{r30}$.

Why is pose important for robotics? It is not as important for static robots, such as stationary robotic arms, though it can be used to indicate the direction the arm is facing at any time. However, for mobile robots, it becomes very important. The robot needs to navigate through a space. To do this, it can perform map-based navigation and reactive navigation. Map-based navigation uses a map based on the frame that indicates to the robot the static elements of the environment. If nothing other than robots exist in the space and the map is updated anytime any robot moves a static object, map-based navigation may be enough. Reactive navigation is based on local robot sensors that detect obstacles and the robot navigates around them. Map-based and reactive navigation can be used together.

As an additional level, robotics pose processing often includes multiple frames. There is the "world" frame or "building" frame and then the robot frame at a minimum. The robot frame is based on the location of other items in the space from the perspective of the robot, which offsets the world frame to the position of the robot. Additionally, some number of other robot frames may also be available when the robots communicate with one another. That is, robot A can inform robot B of its location and frame in relation to an object and the two robots, through information sharing, can each more accurately locate each other and other objects.

Frames include the global coordinate system (GCS), the machine coordinate system (MCS), the tool coordinate system (TCS), and the workspace coordinate system (WCS). The GCS is the "world" frame. It defines where the machine or robot is in the space. The MCS is the machine or robot perspective frame. The TCS is the tool perspective frame. A single robot may have multiple tools, and each would have a TCS. The WCS is the frame of the workspace in which the robot may operate.

As you can imagine, the mathematics behind multiple frames becomes quite complicated quickly in 2D space and 3D frames become even more complicated. Add to this the additional concept of SLAM (simultaneous localization and mapping), wherein robots

> build a map using range sensors as they move throughout a space, and you can see that modern robotics has come a very long way.
>
> Today, the vast majority of robots in use are stationary. They fall into the cartesian, cylindrical, spherical, articulated (anthropomorphic), puma, and scara categories. These categories simply define the movement of the robots within stationary space.
>
> *-Tom*

Bluetooth Low Energy (BLE) Localization (Direction-Finding)

BLE direction-finding is based on angle-of-arrival (AoA) or angle-of-departure (AoD). The complex mathematical algorithms are not necessary for understanding the design issues involved in implementing BLE localization or direction-finding. However, you should understand the differences between the two methods and what is required for them to work properly.

AoA is used by a central device or devices performing the direction-finding (locating another device) and it is based on the received signals from the end devices. The direction-finding device has multiple antennas and uses these different receiving antennas to identify the different characteristics of the signal received at them to determine the direction from which the signal must have originated. In this case, the central device is the locator. More complex architectures allow for multiple locators to share information for even more accurate positioning. Figure 6.9 shows both AoA and AoD processes.

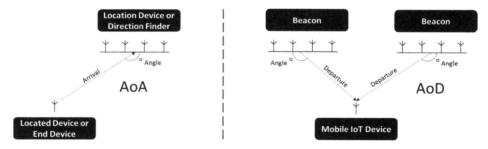

Figure 6.9: BLE AoA and AoD

AoD is used by the mobile or IoT devices to determine their location in relation to beacons. The beacons transmit a signal but also include absolute coordinates in relation to a shared map or floorplan so that the mobile device can determine its location in relation to the multiple beacons it can hear[151].

BLE location services effectively use a broadcast topology, according to the Bluetooth SIG. They suggest that three topologies are available with BLE:

- PtP: Data transfer directly between two devices (watch and mobile phone, for example)
- Broadcast: Locationing and other "announcement" services
- Mesh: Device networks such as IoT end devices

Wi-Fi Localization

Wi-Fi localization is based on a centralized Indoor Positioning System (IPS) or Real-Time Location System (RTLS) that receives signal information about an end device from either 802.11 access points (APs) or dedicated 802.11 sensors. When a device transmits, all APs or sensors that hear it send the signal strength information to the centralized system. The centralized system calculates the position based on known information about the location of each AP/sensor and calculated information about the determined location of the transmitting end device. Figure 6.10 shows this concept.

When using an AP-based solution, the benefit is that sensors are not required so costs are reduced. However, the benefit of a sensor-based solution is that the APs do not have to take time to process the localization information and can dedicate their time to serving client devices. Additionally, APs are normally placed where they best serve the data transmission needs of the client devices and not where they best serve localization. While a deployment plan could be optimized to serve both well, it is more challenging to develop a plan that serves both as well as it could service one function. However, in many implementations, the impact on localization accuracy is not significant enough to

---

[151] To be clear, the BLE specification 5.1 added the more advanced direction-finding capabilities represented here. Before this, BLE could be used for proximity (the estimate of distance between two devices), which was mostly used as locator services, such as finding keys that have a tag on them using a mobile phone. It was also used for positioning (triangulation using three or more locators), where the locators worked together to formulate a position within 1 to 10 meters of the actual location. The accuracy of location was heavily dependent on the quality of the location map and the number of locators available. AoA and AoD make accuracy at the centimeter level possible, taking BLE localization to the next level.

warrant a sensor overlay solution and the performance impact may be minimal as well depending on the solution.

The accuracy of Wi-Fi localization varies. Older PHY technologies tend to be less accurate than newer technologies. However, the real key to location accuracy is often in the vendor secret sauce (the internal algorithms in the IPS or RTLS that determine the location).

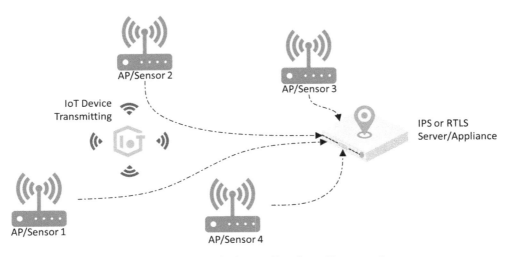

Figure 6.10: Wi-Fi Localization Illustrated

Additionally, different methods are available (or may soon be) for Wi-Fi localization including:

- **RSSI Trilateration:** This algorithm uses three readings of the signal from the target device at three locators. The readings of RSSI are used to calculate three circles (based on the signal strength received and the attenuation over distance, an estimate of distance can be calculated) with the distance travelled being the radius of each circle. They target device location should be approximately where the three circles intersect. Alternatively, if the transmitter and receiver are synchronized, time of arrival (ToA) may be used to determine the distance traveled (given an assumed speed of travel), from which the circles can also be calculated.

- **RSSI Multilateration:** Instead of ToA, multilateration uses Time Difference of Arrival (TDoA) to calculate the distances and is most accurate when four or more receivers are involved in the location process. While it can provide tolerable locationing with three receivers, it is highly accurate with four receivers and for each additional receiver a marginal level of accuracy increases.
- **RSSI Fingerprinting:** This method requires training to generate fingerprints (a probability distribution of RSSI values for a given x,y location) and then can be used for localization.
- **Angle of Arrival (AoA):** Similar to BLE, Wi-Fi devices can potentially perform AoA direction-finding. The availability of such a solution would be vendor specific.
- **Fine Timing Measurement (FTM):** Added to the 802.11 protocol with the 802.11-2016 rollup, this feature allows stations to use Fine Timing Measurement frames to accurately measure round trip time. Using multiple frames, a station can determine its relative location to other stations. If supported through a higher layer application, this can result in 1–2-meter location accuracy.

RFID Localization

RFID (Radio Frequency ID) uses tags that are read by readers when the tags are in proximity of the readers. RFID can be short- or long-range[152]. In short-range RFID the signal is effectively radiated by the equivalent of a printed circuit board (PCB) antenna. In long-range RFID[153], a dipole antenna is typically used for signal transmission. In either case, the simplest localization method used in RFID is the fact that a given reader has just received a signal from the RFID tag (it has read it) and, therefore, the tag must be in close proximity to the reader. For short-range RFID, the tag is typically within 0-30 (0-9 meters) feet when it can be read by the reader[154]. For long-range RFID, the tag may be up to 300 feet (100 meters) from the reader, and, in such cases, trilateration may be used with multiple readers to determine location. With a single reader only, the accuracy of

---

[152] Not to be confused with near-field vs. far-field. Near-field is practical touch reading. Far-field is anything at any separated distance longer than the wavelength. For 960 MHz, for example, this would be more than 32 centimeters.

[153] Long-range RFID is very different from long-range IoT protocols. The former is measured in meters and the latter is measured in kilometers.

[154] The accuracy of localization also depends on the type of RFID tag. RFID tags can be active or passive. They can be HF or UHF. All these factors impact the range of possible communications and therefore the ranging for localization.

localization is limited. With multiple readers, location can be determined within a one-meter range.

RFID localization is often used to simply determine that a tracked physical entity is "here" as opposed to "not here." For example, in inventory management, the concern may not be the location of the package within meters or centimeters, but rather only that the package has arrived in the facility (currently seen) or departed from the facility (no longer seen) [155].

RFID is usually a sub-1 GHz technology operating, typically, from 860 MHz to 960 MHz (UHF), 433 MHz (VHF), 13 MHz (HF), and 134 KHz (LF), depending on the regulatory domain and application. The kind of tag and antenna used with RFID is a factor of cost and tracked asset value or type. For example, a $2 (US) item is not likely to be tracked with a $15 (US) RFID tag and antenna pairing. However, spending $40 (US) for a tracking tag that is applied to a $25,000 (US) device is certainly cost-beneficial. However, in modern IoT, it is very likely that more information than that provided by an RFID tag would be desired with such an expensive device and, therefore, it is likely that alternate IoT sensors/actuators would be used with more "communicative" protocols, which can also track location.

When selecting RFID tags, consider the following factors:

- **Frequency Band:** Must comply with the regulatory domain and provide the desired functionality with tags that operate in the selected band.
- **Use Cases:** Harsh environments may require protective enclosures. Tags within packaging should be near the external of the package (preferably securely attached without). Tags on metal or reflective objects may pose problems and should be tested. This issue is also true when used with water-containing items.
- **Tag Size:** Larger tags are often used with shipping containers, large vehicles, and other large items offering dipole antennas and increased capability. Smaller tags are used for merchandise tagging, pharmaceutical tagging, and other small form-factor applications and are typically less capable.

---

[155] RFID is primarily focused on its namesake: identification. Using RFID for location tracking or to provide other information is possible within limits; however, it is still mostly used for its original purpose: identifying an object. For example, identifying an object in inventory, a drug in a hospital, and a tool in manufacturing. The most common additional use of RFID is localization.

- **Storage Capacity:** RFID tags can store as little as 8- or 16-bits of data to as much as 64 kilobytes of data. Tags supporting storage of the required information should be chosen. In some solutions, the tags can have a small identifier and everything about the physical object is stored within a distributed system associated with that identifier. In other solutions, where physical objects are received from other organizations and no integration exists with that other organization's information systems, more information may be required on the tags to provide sufficient data for processing.
- **Mounting Methods:** Methods include cable ties, tape, glue, screws, rivets, welding, embedding and more. Choose a method that ensures reliability in maintaining the RFID tag with the associated physical objects. In some cases, placing it in a box with other items is all that's required. In others, it must be securely mounted, which is particularly common when the RFID tag is more permanent than temporary.

Beyond the Exam: Is RFID IoT?

People sometimes ask if RFID is an IoT protocol. A simple response would be that it is the original IoT protocol. When Kevin Ashton coined the phrase Internet of Things, it was in relation to an RFID project. Interestingly, he only called it Internet of Things because the term *Internet* was a buzzword in the late 1990s and he felt it would get the attention of the executives. It was not because Internet connectivity was a primary focus of the technology.

According to Kevin Ashton[156], he titled a presentation *Internet of Things* because the Internet was the popular concept all executives wanted to implement, capitalize on, and utilize in 1999. He was working with Proctor & Gamble at the time and the project was an RFID project in supply chain optimization.

Apparently, he had tried without success for six months prior to the presentation to convince executives that putting RFID tags on products in the supply chain would be beneficial for location tracking, inventory management, and sales tracking. Finally, he had to give a presentation that would change their minds. He was going to discuss the concept of a network of things and the Internet being a network of bits and how sensors

---

[156] https://iot-analytics.com/internet-of-things-definition/

> would bring them together and he said, *"Then I thought of an 'Internet of Things', and I thought, 'That'll do – or maybe even better.' It had a ring to it. It became the title of the presentation."*[157]
>
> The rest is history. Kevin moved on to co-found the Auto-ID Center at MIT, which was a primary research lab that helped to build the foundation of the modern Internet of Things. So, is RFID IoT? More accurately, RFID was the genesis of IoT, and it would seem rather odd to rule it out now.
>
> -Tom

Protocols and Channels

The various wireless IoT protocols operate in different frequency bands. These frequency bands, in part, assist in accommodating the intended us of the protocol. For example, long-range protocols tend to use lower frequency bands. Short-range protocols tend to use higher frequency bands. While additional factors impact the range of a protocol, the frequencies used for channelization have a significant impact. Figure 6.11 illustrates the common IoT frequency ranges used and the characteristics of those ranges.

400 MHz	**Very Long Range**	3.5 GHz	**CBRS**
	Low data rates		Low to medium data rates
	Resiliency and very long range		LTE-like resiliency and range
	Smallest number of protocols		Unique use cases
800-900 MHz	**Long Range**	5 GHz	**Wi-Fi**
	Low data rates		Mostly used by Wi-Fi
	Resiliency and long range		High data rates
	Medium number of protocols		Short range
2.4 GHz	**Worldwide**	6 GHz	**Wi-Fi**
	Medium to high data rates		Mostly used by Wi-Fi
	Heavy contention		High data rates
	Medium range		High resiliency and efficiency
	Largest number of protocols		Short range

Figure 6.11: Frequency Band/Range Characteristics

---

[157] https://blog.avast.com/kevin-ashton-named-the-internet-of-things

The various protocols also match different characteristics of importance during selection for IoT solutions. These are depicted in Figure 6.12 and include:

- **Range**[158]: In the image, closer to 10 indicates greater range and closer to 0 indicates lesser range. All three long-range protocols are similar in range; however, variation exists between the IoT protocols, with the latest Bluetooth offerings granting the greatest potential range, though few systems have taken advantage of it at this time. The comparisons are based on like hardware (antennas, transmit power, etc.) used for the protocols. Certainly, any protocol can accomplish greater range with high gain antennas and the highest allowed output power, but when you compare BLE coded PHY with the lowest Wi-Fi data rate on the exact same hardware with the same transmit power, BLE has a longer range.
- **Energy Management:** For this metric, higher values (closer to 10) indicate less energy consumption and lower values indicate more energy consumption. The long-range protocols are all very close, but Wi-Fi generally consumes the most power in the short-range protocols listed.
- **Data Rate:** Higher values for data rate indicate higher data rates. The long-range protocols are all very low in data rates. Wi-Fi has the highest data rates and Zigbee has the lowest in the listed short-range protocols. However, having a higher data rate is not always (or even commonly) the goal in IoT. The goal is to have a consistently stable link that provides the throughput required for the communications. Given this real-world requirement, lower data rate protocols can be a better choice in many scenarios.
- **Latency:** Latency is a more complex metric in this case. It is not intended to be latency in the duration required to send a bit on the medium. Instead, it is a measure of internal and external latency. Higher values indicate that the data gets to the destination faster with that protocol and lower values indicate that the data gets there slower.

---

[158] The range of a protocol is also significantly impacted by the modulation used, with more complex modulation functioning at shorter distances but offering higher data rates and less complex modulation functioning at longer distances with lower data rates. An additional factor impacting the range of the protocol is the coding used. Coding methods that tend to result in lower bit throughput (for upper layer payloads) generally allow for longer distances. Finally, the channel bandwidth impacts the range as well, with narrower channels offering greater distance.

While several other factors and protocols could be evaluated, the point of Figure 6.12 is to illustrate most clearly that there are significant differences among the protocols and these differences must be factor into the decision and selection process.

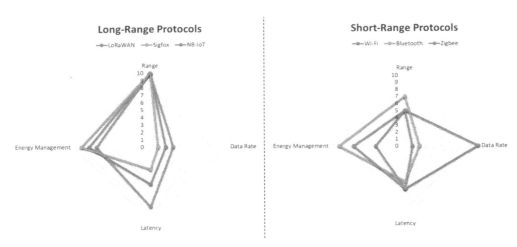

Figure 6.12: IoT Protocol Characteristics

Knowing the frequency bands within which the protocols operate is important and Figure 6.13 shows many common wireless IoT protocols and the most common frequency bands in which they operate. After understanding the frequency bands, you can explore the channels they use. Some device protocols do not use a "channel" per se, but rather communicate using some defined bandwidth within the frequency range allows. The base station listens across the entire frequency range and can receive communications anywhere in that range. This is a very different concept of channelization than used by most protocols and it is implemented by the Sigfox protocol.

The channelization for the protocols listed in Figure 6.13 are defined in Table 6.2[159]. The wireless IoT designer must select channels that are best for use in the target environment.

---

[159] Standards documents do not always directly define a specific channel bandwidth in MHz, but they will indicate the channel bandwidth. For example, the O-QPSK PHY (used by most 802.15.4-based protocols operating in 2.4 GHz) specifies that *the reference level shall be the highest average spectral power measured within ± 1 MHz of the carrier frequency* (the center frequency). That is 2 MHz of bandwidth.

In many cases, the design can specify exact channels to use. In other cases, the system selects the channels automatically and may allow the designer to disallow some channels for optimum configuration.

Non-Cellular	433	470	868	915	920	2450	3500	5800	6000
Bluetooth						◊			
Thread						◊			
6LoWPAN						◊			
LoRaWAN	◊	◊	◊	◊					
Sigfox			◊	◊					
ISA100.11a						◊			
WirelessHART						◊			
Zigbee			◊	◊	◊	◊			
Z-Wave			◊	◊	◊				
802.15.4	◊	◊	◊	◊	◊	◊	◊	◊	
Wi-Fi						◊		◊	◊
CBRS							◊		

Figure 6.13: Most Common Frequency Bands for Popular Wireless IoT Protocols

Protocol	Number of Channels (Max)	Approx. Channel Width
Wi-Fi	98	20 MHz
Bluetooth (BLE)	40 (37 for data)	2 MHz
Zigbee (802.15.4)	16	2 MHz
ISA100.11a (802.15.4)	16	2 MHz
WirelessHART (802.15.4)	15	2 MHz
6LoWPAN/Thread (802.15.4)	16	2 MHz
LoRaWAN	80 (regulatory dependent)	125, 250, and 500 kHz (regulatory dependent)
Sigfox	Dynamic by regulatory	100 Hz or 600 Hz

Table 6.2: Common Wireless IoT Protocols and Channelization

As a side note, many whitepapers, research articles, books, and other publications state that ISA100.11a and WirelessHART use the DSSS PHY from 802.15.4. The problem is that the listing of PHYs does not provide a PHY simply called DSSS. However, in the standard both the O-QPSK PHY and the BPSK PHY are defined as DSSS PHYs and this seems to be where the confusion comes in. The O-QPSK PHY is defined as a *DSSS PHY employing O-QPSK modulation* and the BPSK PHY is defined as a *DSSS PHY employing BPSK modulation*. So, as you can see, a DSSS PHY does exist, it is just not so named.

Channel Selection for Sub-1 GHz Protocols

Sub-1GHz PHYs use license free ISM Bands defined by the International Telecommunication Union (ITU) with 3 main regions:

- ITU Region 1: The European Telecommunications Standards Institute (ETSI) defines the 868 MHz as the main ISM Band used in Europe, Africa and the Middle East.
- ITU Region 2: The US Federal Communications Commission (FCC) defines the 915 MHz as the main ISM Band used in North and South America.
- ITU Region 3: Different Local Regulations and Countries in the Asia-Pacific region have a mix of 868 MHz and 915 MHz bands defined.

In the case of LoRa, the following Channels can be selected based on the region:

- The US915 ISM Band:
  - 64 uplink (upstream) channels numbered 0 to 63 with 125 kHz BW separated by 200 kHz from 902.3 to 914.9 MHz
  - An additional 8 uplink (upstream) channels numbered 64 to 71 with 500 kHz BW separated by 1.6 MHz overlap between 903.0 to 914.2 MHz
  - The downlink (downstream) has 8 Channels numbered 0 to 7 with 500 kHz BW spaced out by 600 kHz between 923.3 and 927.5
- The EU868 ISM Band:
  - Channels must be defined by default but up to 16 channels can be selected with a of bandwidth 125 kHz.
- For other regions please refer to the document available at the LoRa Alliance website titled, "LoRaWAN Regional Parameters."

For Sigfox, there's only one Macro-Channel using 192 kHz of Bandwidth with 320 sub-carriers 600 Hz wide each for Downlink, but for Uplink, there's two scenarios:

- Africa, Asia, and Europe (RC1, 3, 5, 6 and 7): Same number of Macro-Channel as Downlink, but with 1920 sub-carriers 100 Hz wide each.
- The Americas (RC2, and RC4): Six (6) Micro-Channels using 25 kHz each with 250 sub-carriers 600 Hz wide.

For Z-Wave the Channel specifications are defined in the ITU-T G.9959 Standard, most regions support 1 or 2 channel configurations, but 3 Channels are defined in total.

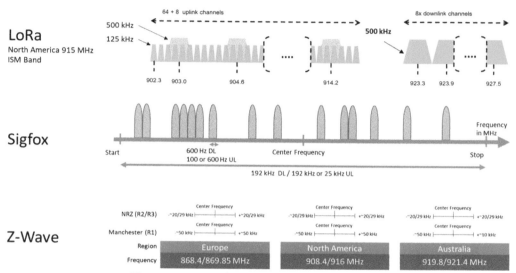

Figure 6.13a: Sub-1 GHz Protocol Channels

Channel Selection for 802.15.4-Based Protocols

The most used channels in the 802.15.4 PHY are defined as follows:

- Only one channel in the 868 MHz band using 300 kHz channel width, mainly for Europe
- Ten channels in the 915 MHz band using 600 kHz channel width and 2 MHz spacing, for the Americas, Asia and Australia.
- Sixteen channels in the 2.4 GHz band with 2 MHz channel width and 5 MHz spacing for Worldwide usage.

The 2.4 GHz band is widely used in many 802.15.4 based technologies like Zigbee, Thread, ISA100.11a, and WirelessHART (with channel 26 disallowed for this last one).

As stated, 802.15.4 IoT operates mostly in the 2.4 GHz band as that is the band primarily used by ISA100.11a, WirelessHART, 6LoWPAN, Thread, and modern Zigbee. This band is quite cluttered in many environments, unless you are designing a wireless network in a rural area with hundreds of meters between you and the nearest neighbor. In the latter case, the band is likely mostly in your control. In other cases, careful planning must precede implementation of any wireless networks in the band.

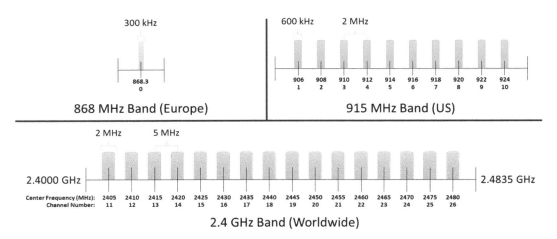

Figure 6.13b: 802.15.4-Based Protocol Channels

The following protocols and systems all commonly compete for time in the 2.4 GHz band:

- 802.11 Wi-Fi
- Bluetooth
- 802.15.4-based Protocols
- Proprietary systems (including phones, cameras, etc.)
- Consumer-grade devices
- Microwave ovens

Just consider Figure 6.14b and the channel possibilities for the shown protocols. Keep in mind that Bluetooth Classic and LE use frequency hopping, so they are all over the band. Traditionally, we have said that it's not a problem because Bluetooth uses low transmit power (compared to Wi-Fi and 802.15.4) and low gain antennas for near-range communications. This fact was true when Bluetooth was used as a peripheral connection protocol. Bluetooth Classic is still primarily used for that. However, BLE is used much differently. It is used for localization and will be used more and more for IoT data transmissions, which can range beyond one kilometer with the coded PHYs. That changes everything. Imagine you have one hundred BLE end devices using the 2 Mbps PHY in an area covering 2500 square meters (50 x 50 meters). BLE's LE 2M PHY can easily operate in that range. Additionally, it can transmit at up to 20 dBm or 100 mW. Now, you have two hundred hoppers moving all over the 2.4 GHz band consuming airtime. Yes, they will use AFH. Yes, they will likely send less data than a Wi-Fi laptop. But, yes, they will be far more problematic in the 2.4 GHz band than the older Bluetooth devices (headsets, mics, etc.) were.

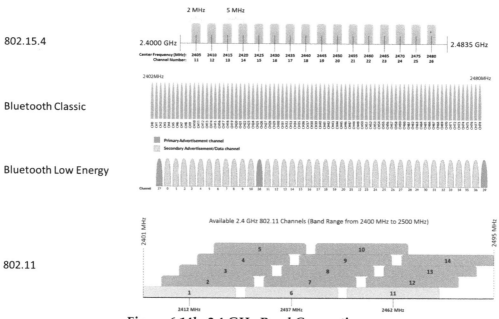

Figure 6.14b: 2.4 GHz Band Congestion

Now, add onto this the possibility of ISA100.11a and/or WirelessHART (802.15.4 PHYs) operating in the same space as BLE IoT. You now have multiple 2.4 GHz protocols that are hoppers. While the 802.15.4 PHYs may allow you to block channels to limit the hopping sequences, the BLE devices do not always offer this though it is available from the standard[160]. They all may use AFH (Adaptive Frequency Hopping) as long as it is enabled, which is all about ensuring good BLE communications. It is not about playing nice with other protocols, do not be confused. It is about the BLE protocol working. It only stops using a channel when the BLE data is not getting through, or the channel is not clear sufficient enough time for transmissions. This functionality is not the same as "playing nice". It is a selfish algorithm, which is not necessarily a bad thing. Much in human psychology today is based on selfish algorithms that result in the ultimate good[161]. Bluetooth is like that, and the problem is that AFH is constrained. That is, a minimum number of channels must be retained regardless of whether they are reported good or bad. This minimum number of channels is 20 and the standard makes no different in this between Classic and LE.

AFH, reporting on channel quality, reports channel health as *unknown*, *good*, or *bad*. Channels reported as bad are blocked from use by the hopping pattern until they are reported good again. Even this function could potentially be problematic depending on how it is implemented as the flipping of channels on and off (because they are good one minute and bad the next) introduces volatility to the band. Also, the standard specifies that AFH shall be disabled by default in that it requires the AFH_mode parameter be set to AFH_disabled unless it is negotiated to be enabled. This default means that the BLE IoT network must explicitly enable and negotiate the use of AFH. It is important then, if

---

[160] The standard allows the master to send a channel list to the slaves that limits the channels used even before AFH does its work. Therefore, the allowed channels can be configured in the BLE gateway and pushed to the end devices.

[161] Hobbes was an early proponent of this in *Leviathan: Or, The Matter, Forme & Power of A Commonwealth Ecclesiasticall and Civil*, in which he stated, "Whensoever a man Transferreth his Right, or Renounceth it ... it is a voluntary act: and of the voluntary acts of every man, the object is some Good to himself." While some modern philosophers doubt or question the theory of selfishness or self-interest as the primary motivator (Miller, 2001; Etzioni, 1988; Mansbridge, 1994; Batson, 1991; and Lerner, 1980, as a few examples), one could question the motive for questioning the motive. That is, does one gain attention in publication by presenting a counter-theory? Also, those who question the motive by suggesting that the existence of the theory since Hobbes has resulted in a self-fulfilling prophecy of people creating institutions and societies that reflect the theory because they accept the theory (Millar, 2001) fail to explain how the same institutions and society constructs existed at various times and in various cultures that never considered the theory. Regardless of where one stands on this philosophical argument, if people act primarily in self-interest, given the much good that does exist, acting in self-interest must create good for others. Therefore, AFH is self-interested (protecting the BLE transmissions), but can result in some positive impact on other protocols in the same area, much as human self-interest benefits others.

using a BLE gateway, to ensure that it is enabled in the configuration, when channel blocking is not available. When channel blocking is available, strategic planning of BLE channels with other systems (Wi-Fi, 802.15.4, etc.) in the airspace, should be performed and appropriate channels blocked and AFH should be enabled as well.

Additional Channel Factors

The following additional channel factors should be considered:

- **Channel Selection:** Select the best channel based on other activity in the area. If the area is free of activity in the entire band used by the protocol, you can use a channel plan throughout your network that simply staggers the channels among different gateways to improve airtime availability. If other systems used the same frequency band, you must select channels near those areas or in the same areas that conflict the least. It is not always possible to select a completely "clean" channel.
- **Channel Bandwidth:** Some protocols, such as LoRaWAN, allow you to configure the channel bandwidth or, depending on the system, it may be selected automatically based on link quality. Generally speaking, with IoT, using narrower channels (less bandwidth) results in lower data rates but more efficient reuse when multiple gateways are required.
- **Dwell Time and Hopping Patterns:** Protocols like ISA100.11a and WirelessHART use channel hopping sequences. The dwell time may be configurable, but the hopping sequence nearly always is. At the very least you can perform channel blocking (we call it blocklist or blocked channels, the standard calls it blacklisting), as it's defined in the standards, which simply means that you are disallowing those channels from use in the hopping sequence or pattern. Bluetooth also uses frequency hopping, but it performs adaptive frequency hopping (AFH) and automatically disallows use of channels on which messages failed to be delivered.
- **Modulation and Coding:** Bluetooth and LoRaWAN allow for changes in modulation and coding depending on link quality or configuration. 802.15.4 protocols, covered in this material, do not allow such adjustments, in most cases they simply use O-QPSK at 250 kbps or they don't work. Bluetooth supports varied coding (how the data bits are converted to bits for transmission) and modulation (how the coded bits are imposed on the signal) as does LoRaWAN, which can increase range. For example, the Bluetooth coded PHY is required for

the long range links the Bluetooth SIG speaks so often about. These PHYs have lower data rates than the uncoded PHYs.

Protocols and Tx Power

Transmit power (Tx power) is an important component in designing any wireless network. The primary purpose of Tx power adjustment is management of cell size and reduction of interference among cells. The cell should be large enough to achieve the required range of all end devices needing to participate in the cell.

The concept of desired range for a given protocol can be summarized as[162].:

$$\text{Range } (\geq dBm_a) = f(Tx_p \text{ \& } F \text{ \& } Tx_g \text{ \& } Rx_g)$$

Where a is the desired received signal strength in dBm, Tx_p is the transmit power in dBm, F is the frequency in MHz, Tx_g is the transmitter antenna gain, and Rx_g is the receiver antenna gain. This logical formula states simply that the range of a transmission for a given protocol with a desired signal strength at the receiver of greater than or equal to the dBm value of a is a function of transmit power, frequency, transmitter gain, and receiver gain. It is for this reason that transmit (Tx) power is so important in wireless network design.

A common problem is the use of transmit power at levels higher than desired for a design. Less common, but also a problem is using transmit power at levels lower than desired for a design. Like Goldilocks, we want the transmit power that is just right. The conceptual range formula provided above is the basic guide.

IoT protocols rarely demand high data rates. Those that do, will likely you Wi-Fi or CBRS or possibly LTE/5G. The other common protocols usually offer 2 Mbps or lower data rates and their selection is not based on achieving high data rates. Their selection is based on achieving resiliency and consistency in low throughput wireless links. Therefore, the goal is to have a link that provides a tolerant and stable connection. Tolerant, because it

---

[162] While the logical conjunction operator "∧" could have been used, the alternate logical operator "&" is easier for the non-mathematicians and non-logicians to process. This range calculation assumes the channel bandwidth, modulation, coding, etc. are constrained because a protocol has been selected.

can continue to function in the presence of noise and other possible interferes. Consistency, because, well, it just works most of the time.

So, why is higher transmit power bad? If you have several hundred or several thousand IoT devices in your network, you are likely to have multiple gateways. The largest of IoT networks will often have dozens or even hundreds of gateways. Generally speaking, for a given coverage area, the more gateways you have, the lower you want the transmit power to be. This allows the gateway to communicate with the devices connected or communicating with it, but it does not cause the gateway to impose unnecessary interference on other gateways and end devices that may be using the same channel at some distance.

Some devices, both gateways and end devices, will allow you to adjust the transmit power. Others will dynamically adjust the transmit power based on the signal they receive from the gateway or the device. Still others will be locked on a transmit power level. In all cases, devices should be operating within regulatory guidelines. However, just because the regulatory guidelines allow you to transmit at 25 mW doesn't mean you want to transmit at 25 mW. The basic rule is to transmit at the power level required to provide a stable link at the desired data rate. This does not mean to transmit at the lowest possible power level that will establish a link. Such a link may not be stable. Some padding for variation in the channel should be accommodated.

Protocols and Data Rates

The common wireless IoT protocols support varying data rates ranging from approximately 100 bits per second (bps) to 9.6 gigabits per second (Gbps). With such a varied range of bit delivery rates, clearly selection of the right protocol is important. When considering the data rate factor in protocol selection, the designer must know the end devices and applications to be used so that a protocol meeting their requirements is chosen. Table 6.3 lists common wireless IoT protocols and their possible data rates.

In nearly all cases with heavy traffic, such as continual video streaming, Wi-Fi or another protocol with data rates higher than a few Mbps will be required. However, these cases do not comprise a large percentage of IoT applications. The vast majority of IoT applications work very well with data rates lower than 1 Mbps. This is because the data payload is quite small. Even if transmissions occur several times per minute, if the entire airtime for each transmission is 30 ms or less (which is quite a lot of time for the indoor, short-range protocols[163]), that means approximately 25-30 such messages could be transmitted each second.

Protocol	Data Rate(s)
Wi-Fi	400 Kbps – 9607.8 Mbps
Bluetooth (BLE)	125 kbps – 3 Mbps
Zigbee (802.15.4)	250 kbps
ISA100.11a (802.15.4)	250 kbps
WirelessHART (802.15.4)	250 kbps
6LoWPAN/Thread (802.15.4)	250 kbps
LoRaWAN	250 bps – 50 kbps (with FSK)
Sigfox	100 – 600 bps
Z-Wave	9.6 – 100 kbps

Table 6.3: Common Wireless IoT Protocols and Data Rates

Capacity is where understanding the data rate and how it impacts airtime utilization becomes an important factor. If you can transmit 30 messages per second across all devices within a channel and you require that 100 devices transmit one message every second, you will need four operating channels for the 100 devices.

More information on capacity calculations is provided in the later section titled *Protocols and Coverage and Capacity*.

To show that lower data rates are better when you want to ensure delivery of a message and speed is not as important, consider Figure 6.14. Here, we show the successful

---

[163] As an example, at 100 kbps, you could send a total bit set, including framing and PHY header information, of around 250 bytes (much larger than most IoT transmission), in less than 30 ms. Given that IoT transmissions are more often 130-150 bytes or less, most transmission consume less than 20 ms, even at 100 kbps. Of course, the long-range protocols, like LoRaWAN and Sigfox, can consume several hundred milliseconds of airtime with the lowest data rates; however, they also constrain the use of airtime in many regulatory domains.

delivery rate and different SNR levels for BPSK modulation versus 64-QAM modulation[164]. The leftmost boxes in the image represent the intended symbols. The rightmost three boxes in each row represent the received symbols.

The image shows the percent of successfully demodulated symbols for BPSK compared to 64-QAM at the same SNR levels using constellation charts to illustrate it. This image reveals why the IoT protocols can be more resilient than high data rate protocols like Wi-Fi. First, they often use narrower channels, which reduces noise in the receiver. Second, they can differentiate between signal and noise more easily due to the simpler modulation. The low data rates are not a problem for most IoT scenarios. When an IoT scenario demands high data rates, the most common options are Wi-Fi, LTE/4G or 5G, or proprietary solutions, which are often still just OFDM under-the-hood, like Wi-Fi.

In the image, you can see that the intended symbols consist of only two possibilities for BPSK and 64 possibilities for 64-QAM. This means that BPSK can represent one of two values in each symbol (1 bit per symbol) and 64-QAM can represent one of 64 values in each symbol (6 bits per symbol[165]), giving 64-QAM a much higher data rate. However, in the tests performed by Calebaut, et. al., only 18% of messages were delivered using 64-QAM when the SNR as low at 6 dB. Compare that to 99.5% delivery when BPSK is used at the same SNR.

The point is that high data rate, and therefore potentially high throughput, protocols require significantly higher SNR to demodulate the transmissions. Low data rate, and therefore low throughput, protocols often allow for negative SNR while maintaining sufficient link quality for communications. The data in Figure 6.14 is based on the link quality specification for LTE of 20 dB as a good link and this is the purpose for the 20 dB beginning comparison. However, while some protocols with varying coding methods may be able to perform slightly better or slightly worse, the general concept applies to all protocols. When more complex modulation is used, a higher SNR is required.

---

[164] Image reproduced and simplified from *The Art of Designing Remote IoT Devices* by Calebaut, et. al., 2021. Sensors journal.

[165] If you are familiar with Wi-Fi, do not confuse the term symbol here with an OFDM symbol in Wi-Fi. The OFDM "symbol" is a reference to all the symbols in all the subcarriers, which is effectively the subcarrier symbol bitrate (6 for 64-QAM) multiplied by the number of data subcarriers (48 in a 20 MHz channel) for a total bitrate of 288 bits per symbols at the "OFDM level."

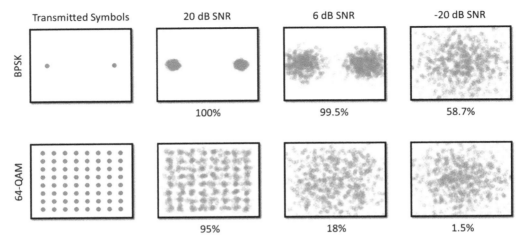

Figure 6.14: Modulation and Delivery Rates

The situation becomes still more interesting when you consider negative SNR values (like those supported by Sigfox and LoRa, though LoRa does not use BPSK). The tests showed that just under 60% of the messages could be properly decoded at a negative SNR value with BPSK but only 1.5% could be properly decoded with 64-QAM. Hence, the general rules of modulation apply:

- the more complex the modulation, the higher the data rate

AND

- the more complex the modulation, the better the quality of link required

Protocols and Uniqueness

Each IoT protocol has some unique characteristics in operation, data transmission, security, and other areas. This section will present some of the important unique characteristics of common wireless IoT protocols. We will include both unique characteristics of the protocols and constraints of the protocols, which will impact your selection decision.

Zigbee

Zigbee is by far the oldest of the protocols referenced here. It has been around since the early 2000s and is based in IEEE 802.15.4. Early Zigbee devices operated in sub-1 GHz and 2.4 GHz, but most modern devices are 2.4 GHz-only. The Zigbee specification states that a network can support up to 65,000 nodes; however, the number of nodes practically supported is a factor of capacity related to the number of channels available for operation. The indoor range is less than 100 meters, and the outdoor range can be longer. Zigbee provides the entire stack for communications within the network and many Zigbee coordinators also act as gateways to provide forwarding of data to business applications. Zigbee is managed by the Connectivity Standards Alliance (CSA) (formerly, the Zigbee Alliance).

6LoWPAN/Thread

The first and most obvious characteristics of 6LoWPAN and Thread networks is support for IPv6, which is the entire purpose of their development. These networks also operate on IEEE 802.15.4 MAC/PHY layers and operate in the 2.4 GHz band. Developments are underway to support them in the sub-1 GHz bands. A Thread network, and most other 6LoWPAN solutions, supports approximately 16,000 end devices per network. Thread, specifically, supports one leader in each network and 32 routers. Each router can support 511 end devices, giving a grand total of 16,351 end devices per network. Of course, 6LoWPAN networks are IPv6 networks, therefore routing between networks is supported and this allows scalability into the hundreds of thousands of end devices. 6LoWPAN is managed by the IETF and Thread is managed by the Thread Group, but Google released an open-source version as the OpenThread standard, which is an implementation of Thread.

ISA100.11a/WirelessHART

Both ISA100.11a and WirelessHART share some commonalities in the lower layers. They are both based on 802.15.4 and implement time-slotted channel hopping (TSCH) and superframes. TSCH provide a mechanism for channel hopping across the 15 (WirelessHART) or 16 (ISA100.11a) channels supported. Superframes are, well, not really frames. They are a method for scheduling communications between peers. Using channel hopping algorithms and superframes together allows multiple communications to occur concurrently (because they are on different channels in the hopping patterns) and ensures device pairs have an opportunity to communicate.

One very important different between ISA100.11a and WirelessHART is how the function at the higher layers. WirelessHART is intended to be a wireless-enabling protocol for HART networks[166]. Therefore, it uses the HART protocols at the upper layers. HART messages are transmitted over wireless links. ISA100.11a is different and, in fact, references RFC 4944[167] as a normative reference in the specification. It is designed to support 6LoWPAN from the start. However, the standard does state the following:

> This standard's network layer uses header formats that are compatible with the IETF's 6LoWPAN standard to facility *potential* use of 6LoWPAN network as backbone. It should be noted that the use by this standard of headers compatible with 6LoWPAN does not imply that the backbone needs to be based on the Internet Protocol (IP). Furthermore, the use of header formats based on 6LoWPAN and IP does not imply that a network based on this standard is open to internet hacking; in fact, networks based on this standard will typically not even be connected to the Internet.

From this statement, you can see that ISA100.11a is designed to support 6LoWPAN but does not require its use on the backbone network. However, the messages sent *within* the ISA100.11a network are indeed IPv6 Network layer packets in many implementations. The standard states clearly that *each device shall also have a 128-bit network address that is assigned by the system manager* and these addresses can either be IPv6 addresses or fieldbus addresses (this address is 128-bits in the ISA100.11a network and is certainly large enough to address communications with the more common 8-bit addresses in direct fieldbus devices).

Bluetooth

The key differentiator with Bluetooth compared to the previously covered protocols in this section is support for higher data rates. All three of the previous protocols have a maximum data rate of 250 kbps whereas Bluetooth (in common IoT implementations) can support data rates from 125 kbps to 2 Mbps. Of course, the extended range supported by newer Bluetooth PHYs (like LE coded) limit the data rate to 125 kbps or 500 kbps. Bluetooth also supports a complete stack through application profiles. However, one of

---

[166] HART is a wired protocol for industrial communications. It defines the command structure and messages used in communications and WirelessHART implements these over a wireless medium (802.15.4) instead of a wired medium.
[167] RFC 4944 is the 6LoWPAN base standard by the IETF.

the features of Bluetooth, specifically BLE, that has driven most of its successes is the integrated support for localization. The newest standards support AoA and AoD for very accurate indoor localization, making Bluetooth an excellent choice when wayfinding, device locating, or other localization needs are defined[168].

LoRaWAN

LoRaWAN is the first of the two long-range protocols covered in this section. It uses LoRa chirp modulation for the primary PHY with support for FSK as well. It supports transmission reception at mare than 20 dB below the noise floor, making it an exceptional long-range protocol. The range is typically defined as between 10 and 15 kilometers; however, this range depends on the terrain and the data rate desired. Depending on the regulatory domain a duty cycle constraint may be imposed. For example, in the EU the allowed duty cycle is 1%. This duty cycle is calculated per hour and it is based on airtime consumption. An excellent airtime calculator for LoRaWAN is available at:

https://avbentem.github.io/airtime-calculator

You can select the regulatory domain under evaluation and then enter the payload size per message to calculate the airtime per message for each LoRa data rage based on various bandwidths and spreading factors. Figure 6.15 shows the tool performing calculations based on an overhead of 13 bytes and a payload size of 10 bytes. You can see that these parameters result in 28.3 ms of airtime utilization for the highest data rate and 370.7 ms for the lowest data rate.

As Figure 6.15 illustrates, the spreading factor (SF) has a significant impact on the data rate and, therefore, airtime utilization. A lower SF value results in a higher data rate for a given bandwidth and a higher SF value results in a lower data rate. To understand this, you must grasp the chirp modulation used by LoRa as presented in Figure 6.16.

---

[168] To be clear, any wireless protocol that provides signal data to the upper layers could be used for localization. However, Bluetooth integrates this into the protocol design making it more appealing to many. Before assuming that Bluetooth must be used for localization and some other protocol for all other IoT operations, check with the possible IoT vendors to verify their localization capabilities.

Figure 6.15: Airtime Calculator for LoRaWAN (US FCC)

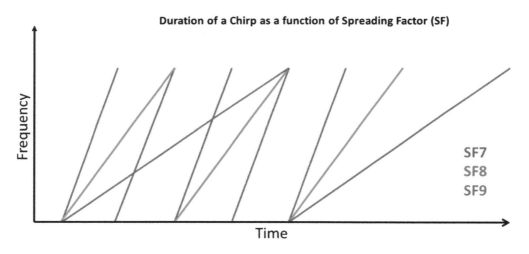

Figure 6.16: Spreading Factor and Chirp Modulation

The image shows the different in chirp spreading in time. If the chip spreading factor is higher, it takes more time for the chirp to complete. Therefore, in a given window of time, fewer chips can be modulated, and the data rate goes down. Table 6.4 shows the spreading factor, SNR limit, receive sensitivity, time on air, symbols per second, and ranges for a 125 kHz channel.

Spreading Factor	SNR Limit	Sensitivity	Time on Air	Symbols/s	Target Range
SF12	-20 dB	-137 dBm	991 ms	30	14 km
SF11	-17.5 dB	-134.5 dBm	577 ms	61	10 km
SF10	-15 dB	-132 dBm	288 ms	122	8 km
SF9	-12.5 dB	-129 dBm	144 ms	244	6 km
SF8	-10 dB	-126 dBm	72 ms	488	4 km
SF7	-7.5 dB	-123 dBm	41 ms	976	2 km

Table 6.4: LoRa Spreading Factor, SNR, and Range

Another unique topic with LoRaWAN is the devices classes, which can be one of three: A, B, and C. These classes determine how the device operates in relation to power management.

- **Class A:** All devices support bi-directional communications such that an uplink transmission from an end-device is always followed by to short downlink receive windows. Class A devices cannot receive communications from the network (downlink) except for the time during the receive windows immediately after an uplink. Downlink communications from the server will have to wait for the next uplink. These devices consume the least power.
- **Class B:** These devices provide more receive windows. The Class A receive process is still supported, but Class B devices open additional receive windows at scheduled times. A time synchronization beacon is sent from the gateway to provide the scheduling for Class B devices. These devices consume moderate power.
- **Class C:** These devices have open receive windows. They are unable to receive when transmitting, but other than that time, they can receive at any time. The radio is always powered on and listening. These devices consume the most power.

Finally, an important consideration is the requirement of a Network ID for any LoRaWAN network implementations wherein you desire for your devices to roam to other network providers. If you want to implement a private network that only your devices can join, but that allows your devices to roam to other providers easily, you will need to register for a Network IT, which incurs a significant annual fee. However, it you wish to implement a private network with your own gateways and servers, you will not need a Network ID. Additionally, you the Things Network provides coverage in your area, and you can operate within the constraints of that network, you can simply use their services without requiring a Network ID.

Sigfox

Like LoRaWAN, Sigfox is a long-range protocol. It also has unique characteristics of importance. First, the specification defines radio configurations (RC1 through RC7) that are used within different regions. The RCs define the center frequencies for uplink and downlink as well as the maximum uplink EIRP and duty cycle or operational requirements as outlines in Table 6.5.

Radio Config	Region / Country	Uplink center frequency (MHz)	Downlink center frequency (MHz)	Max UL EIRP	Regulation Requirements
RC1	Europe, Middle East and Africa	868.130	869.525	16 dBm	1% Duty cycle
RC2	North America + Brazil	902.200	905.200	24 dBm	Frequency hopping
RC3	Japan	923.200	922.200	16 dBm	Listen before talk
RC4	Latin America and Asia Pacific	920.800	922.300	24 dBm	Frequency hopping
RC5	South Korea	923.300	922.300	14 dBm	Listen before talk
RC6	India	865.200	866.300	16 dBm	
RC7	Russia	868.800	869.100	16 dBm	1% Duty cycle

Table 6.5: Sigfox RCs and Regulations

Regulation requirements:

- Duty cycle is 1% of the time per hour (36 seconds). For an 8 to 12 bytes payload, this means 6 messages per hour, 140 per day.

- Frequency hopping: The device broadcasts each message 3 times on 3 different frequencies. Maximum of 400 ms of airtime utilization per channel with no new transmission before 20 seconds have elapsed.
- Listen Before Talk: Devices must verify that the Sigfox-operated 192 kHz channel is free of any signal stronger than −80 dBm before transmitting.

Sigfox's max limit EIRP recommendation is included in each column although regulations sometimes allow for more radiated power than the Sigfox recommendation.

Sigfox uses either GFSK or D-BPSK for modulation achieving data rates between 100 and 600 bps and can support RSSIs as low as -135 dBm as shown in Table 6.6.

Link	Channel Width	Modulation	Data Rates	Target RSSI
DL	600 Hz	GFSK	600 bps	-126 dBm
UL	100 Hz	DBPSK	100 bps	-135 dBm
	600 Hz	DBPSK	600 bps	-127 dBm

Table 6.6: Sigfox Modulation, Data Rates, and Target RSSIs

Cellular and IoT

While cellular is not a primary focus of the CWNP wireless IoT track, it is an important part of many IoT deployments. It is not a primary focus of the track for the simple reason that understanding the inner-workings of cellular is not essential to deployment of cellular-based IoT solutions. One must ensure proper coverage indoors or outdoors, depending on the scenario, and ensure that the service provider provides the required services at the service level required through service level agreements (SLAs). How the service provider "provides" the services, as long as they are functioning according to the agreement, is less important to the implementer.

At times, using cellular may require the implementation of indoor solutions that supplement the carrier's signal; however, such deployments are frequently done with the assistance of the service provider and the design is normally performed by the service provider or in cooperation with them. Such solutions often used distributed antenna systems (DAS) or small cell technologies, which are often designed by service providers or specialty consulting firms[169].

The benefits of cellular communications for IoT projects are several. In many areas, cellular coverage is ubiquitous outdoors. For such outdoor deployments, cellular can be an easy choice if the ongoing budget allows for the service fees. Indoor coverage is not as ubiquitous with many buildings being problematic for signal reception and transmission; however, one service provider may be better than another and this becomes a primary factor in service selection. A final benefit is the common availability of multiple service providers using different networks. This situation can be beneficial when failover to an alternate service provider is desired and is most common when implement IoT for equipment that is used in various locations, such as earth moving equipment and other tools and machines that are used at different sites.

As an example, in the case of large industrial and earth moving equipment, localization is less important[170] than performance metrics. Such large equipment is harder to steal or misplace and this reduces the need for localization, though fifty percent of owners indicate that localization is still important. However, more than ninety percent indicate that performance telemetry is most important. This telemetry data allows them to verify the health of the equipment and perform maintenance as required. It can also improve worker safety by keeping the equipment in optimal operating condition.

Protocols and Coverage and Capacity

Now that we have reviewed the basic characteristics of common protocols, including localization, channels, transmit power, and data rates, we can consider coverage and

---

[169] To be clear, you can certainly design your own DAS solution. It is simply common to outsource it for many organizations because of the lack of internal resources with knowledge of DAS and other cellular technologies. By all means, if you can design and deploy a DAS solution, do it for your organization. However, the primary focus of CWNP's IoT track is on open protocols that operate in license-free bands currently. As IoT evolves, this may change in the future, but that change will occur when and if the experts that assist in designing the certifications see it as essential knowledge for the track. Additionally, small cell solutions are becoming quite popular with 5G and, in appearance, look like Wi-Fi access points deployed throughout the building or coverage area.

[170] www.iotm2mcouncil.org/iot-library/articles/smart-industries/sweating-the-big-stuff/

capacity planning for the wireless network. We will explore this from two perspectives: indoor networks and outdoor networks; however, keep in mind that some networks, such as LoRaWAN networks, may be indoor/outdoor networks.

Wireless IoT Indoor Networks

For indoor networks coverage and capacity, we will use the model represented in Figure 6.17, which addresses coverage morphologies, quality objectives, capacity requirements, propagation models, and interference testing.

The first topic introduced in the model is *coverage morphologies*. Given that this may be a new term to many, let's explore the concept. The term coverage morphology is borrowed from the very technical process of terrain mapping. In these spaces, the terrain is defined as slopes, reliefs, altitudes, and so on. The morphology of the terrain is defined by its critical points and by the critical lines joining them, which form a surface network[171]. The concept of a terrain morphology has evolved into the concept of coverage morphologies in wireless networks and analyzed in several research papers as a solution to coverage problems for Distributes Antenna Systems (DAS)[172], Wireless Sensor Networks (WSNs)[173], and IoT networks[174]. We have adopted the term as a simple way to categorize the different environments in which wireless IoT may be deployed.

While the model represents four environment types, more may be defined. The four environment types are:

- **Office Space:** The space is characterized by humans, office furniture, floor carpeting, drop ceilings, cubicle barriers, walls, elevators, and often multiple floors. The nature of many IoT protocols allows them to work very well in such spaces with fewer gateways as long as the airtime utilization by the devices can be accommodated by a few gateways.

---

[171] Danavaro, Floriani, Papaleo, Vitali. *A Multi-Resolution Representation for Terrain Morphology*. 2006. Lecture Notes in Computer Science.

[172] Xu, Lai, Lang. *A Novel Mathematical Morphology Based Antenna Deployment Scheme for Indoor Wireless Coverage*. 2014. Paper presented at 2014 IEEE 80th Vehicular Technology Conference.

[173] Fu, Yang, Hong, Hou. *WSN Deployment Strategy for Real 3D Terrain Coverage Based on Greedy Algorithm with DEM probability Coverage Model*. 2021. Electronics journal.

[174] Lopez-Iturri, Celaya-Echarri, Azpilicueta, et. al. *Integration of Autonomous Wireless Sensor Networks in Academic School Gardens*. 2018. Sensors journal.

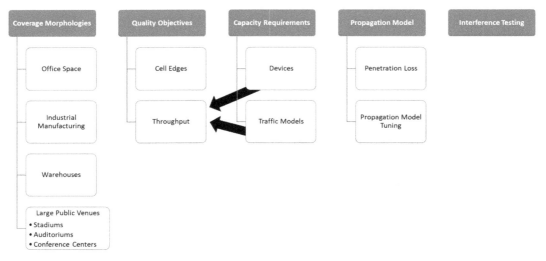

Figure 6.17: Indoor Networks Coverage and Capacity Model[175]

- **Industrial Manufacturing:** The space is characterized by many large metal machines, conveyor belts, moving entities (both humans and machinery), and potential intentional and unintentional RF radiators. Spectrum analysis is nearly always required in such spaces to ensure availability of sufficient uninterfered or low interference frequency space. Protocols and/or channels can be selected to accommodate the space only when the RF activity in the space is known. Additionally, the large metal machinery, some of which is mobile, will alter the propagation patterns from plant-to-plant significantly. Such an environment is a prime target for a Gateway-on-a-Stick (GoS)[176] survey or efficient CW testing.

---

[175] This coverage and capacity model is defined to provide a guide for the issues that must be considered when designing for both in a wireless IoT network. Like any model or framework, additional concerns may be important to consider, and some listed concerns may not be of concern.
[176] Borrowed from the early days of Wi-Fi surveys, we define a GoS survey, as one that uses a gateway that will be implemented in the space to find appropriate locations for the multiple gateways required by placing the gateway in a given area and then evaluating the coverage and then moving it to a new area and so forth until the entire facility is covered. Such a method is far preferred over common vendor recommendations of, "place the gateway in the center of the coverage," leading many to think that every network needs a single gateway, and it should just be placed in the middle.

More gateways may be required in such a deployment than in office spaces due to the nature of the environment.
- **Warehouses:** Warehouses may be similar to industrial manufacturing but differ in a few key areas. First, they are less likely to have as many unintentional radiators (fewer motorized machines in the space) as industrial manufacturing spaces. Second, they have very different propagation impacting components such as large metal shelving and varying inventory loads. The design of the wireless IoT network must accommodate for functionality when the inventory is full and when it is not, and this may require more gateways than are required at all times.
- **Large Public Venues:** Stadiums, auditoriums, and conference centers may seem like environments that would present no challenges to the IoT protocols we are covering; however, they present variability factors that must be considered. The primary variability factors are humans and equipment. In all three, RF propagation through the space will be very different when packed with people than when it is empty. This must be considered when designing the network. In auditoriums and conference centers, the equipment in the space will change as well. However, the good news is that, in such environments, we are rarely providing IoT networks "for" the people but rather for the services we offer: temperature control, lighting control, sound control, traffic monitoring, etc. We are not expecting the people to connect to our IoT network and so we expect to have a lesser impact when people and equipment arrive than would be the case, for example, when providing guest Wi-Fi networks. In many cases, our IoT network can communicate "above the fray." That is, the IoT links can use frequencies different than those used by guest devices and our links can even be established at higher elevations so that human body attenuation and equipment changes have less of an impact. Still, we have plenty of IoT devices that simply must be installed "down with the people" and the change in propagation must be considered in the design[177].

---

[177] Contact tracing or simply people tracing is become more and more common today. In IoT, this can be performed with video-based AI, unique signal tracking (from mobile devices), and many other techniques. The point is that, with this new desire resulting in part from COVID-19 (though it was already in the works, it has simple been accelerated), will have a significant impact on future designs of large public venue IoT deployments.

Knowing the common characteristics of these environments can alert you to frequent issues that will impact coverage within them.

The second topic in the model is *quality objectives*. The quality objectives will feed into the next topic, *capacity requirements*. That is, the capacity requirements of the devices (total number) and traffic models (how much data do the devices transmit and how often) must be accommodated within quality objectives and the quality objectives must be achieved withing device constraints and traffic models. They are interdependent.

The first quality objective is the simplest: you need coverage where you need coverage. This objective is defined through minimum cell edges for the indoor wireless IoT network. Because we wish to meet the second quality objective of throughput, we must also ensure that the cell edges offer signal strength sufficient for the required data rate. However, when throughput is considered, the simple model of "place the gateway at the center of the coverage area" doesn't work. A single gateway may not be able to provide the throughput required due to airtime constraints. Therefore, part of this process is ensuring that sufficient cells are available to allow the throughput required based on capacity demands from devices and traffic models[178].

On the RF side, understanding how the signals will propagate through the indoor space (propagation model) is essential to creating the required cells. This propagation will be impacted by typical RF behaviors you learned about in the CWISA certification. The penetration loss of walls and other obstacles will impact the range of the signals. Propagation modeling can be performed with wireless design software, such as iBwave Design and others. Additionally, they may offer propagation model tunning, which can be achieved using carrier wave (CW) testing. A CW signal generator is placed in the facility and a measurement device is used to determine the signal strength at various locations. The design software can then use the CW testing data to optimize the propagation model for the design project.

Finally, and of great significance, is interference testing. Capacity will be diminished in the presence of interferers. Interferers can cause retransmissions or lost data. It can also cause data rate reduction for protocols that support it. Protocols that use only a defined data rate, like Zigbee, ISA100.11a, WirelessHART, etc., will simply have to retransmit the

---

[178] The number of supported devices is based on several factors: hard limits in gateways (some support as few as a couple dozen devices and some support thousands), airtime constraints, and physical space constraints (you can only fit so many devices in a given 3D space in reality).

data if some mechanism is provided to do so, or the data will be lost[179]. Interference testing can be performed with a spectrum analyzer to locate signals in the space or RF energy in the space that could interfere with transmissions. To design for proper capacity, channels with interferers that cannot be removed or reconfigured should be avoided when possible.

Wireless IoT Outdoor Networks

Coverage and capacity for outdoor networks has some similar topics in the model to those in the indoor networks model; however, some are defined in a different way, and some may be entirely different depending on the type of outdoor network. Figure 6.18 represents the outdoor model.

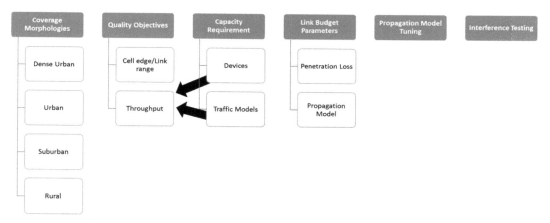

Figure 6.18: Outdoor Networks Coverage and Capacity Model

---

[179] For those coming to the wireless IoT world from Wi-Fi (which is of course also a wireless IoT protocol) it is not always immediately obvious that many IoT protocols have a single data rate and that is it. They do not have the dozens of data rates possible in Wi-Fi networks and therefore to not always have the flexibility to back off the data rate if the link is not good enough for it. Instead, they simply lose connection. Now, the data rate is already low and so they typically achieve this data rate at boundary ranges similar to those of Wi-Fi (for short-range protocols), but this must be considered. The only reason the data rates are achieved at ranges similar to Wi-Fi is the common use of low gain antennas and low output power for short-range protocol implementations. Given the low-complexity modulation and narrower channels used, with higher gain antennas and higher output power, they would achieve ranges significantly longer than typical Wi-Fi. Remember, range is based on the total system from the radios to the antennas and the way the RF is used.

The first noted area of difference is the coverage morphologies. We are now dealing with outdoor environments that may be categorized generically as dense urban, urban, suburban, and rural, borrowed loosely from the cellular space[180].

- **Dense Urban:** Areas are characterized by streets lined with buildings of twenty or more stories resulting in "urban canyons", heavy traffic vehicles of varying sizes, trees along streets impacting propagation. More distributed gateways may be required throughout the environment, or the use of mesh-based solutions can be used. With a mesh solution, each light pole, for example, could have an attached mesh node allowing transmissions to make their way through the environment to some number of distributed gateways.
- **Urban:** Similar to denser urban, but with either fewer buildings or fewer stories in the buildings, slightly wider streets and less traffic. The same design solutions can be used here as in dense urban environments; however, fewer gateways and or mesh nodes are likely to be required.
- **Suburban:** Few buildings above ten stories, buildings are more spaced, many wide-open spaces, but likely with more vegetation including older, taller, and thicker trees. Possibly more rolling hills and less traffic still. Depending on the terrain itself, this type of environment will require even fewer gateways or mesh nodes. The terrain, of course, must be considered in all of the morphologies addressed here.
- **Rural:** Very wide-open spaces, most buildings are one to three stories, very low traffic levels, but may have dense forests depending on the region. These environments may present the least challenges to propagation (though that's not always the case depending on terrain and vegetation), but they can provide significant challenges to design for power provisioning. In the other three morphologies, you can likely power devices through existing AC power at the mount points, if desired. In rural deployments, energy harvesting, and battery power will often become a requirement for functionality[181].

---

[180] Based on ITU-R P.1411-6 and MITRE. *Investigations and Analysis of LTE Network Cell Tower Deployments and Impact on Path Loss Calculations.* 2019.

[181] Using energy harvesting in dense urban, urban, and suburban coverage morphologies is also a common requirement, but it is often less functional and more non-functional. That is, it is required to achieve the quality of less energy consumption from the power grid, which is equally important today.

Certainly, other morphologies such as residential and undeveloped, or mountainous and flatland, or wetland and dryland, and so on could be identified. Over time, with experience, you will begin to discover the impact these different morphologies have on the planning of the wireless links.

With the outdoor environment categorized, the quality and capacity objectives and requirements can be evaluated. They function similar to the indoor environment when implementing cell-based coverage. However, outdoor long-range links are often planned based on each link or each link area. A link area is a sector some distance from the gateway that is the target of coverage. That is, coverage within a few meters of the gateway may not be required at all. However, coverage one kilometer away may be required. In such cases, the designer can choose to mount the antenna attached to the gateway on a pole, roof, or tower with the appropriate angle and downtilt (if required) to achieve coverage at the remote location.

In one project, coverage was desired in a building that was approximately two kilometers away from the source gateway. Rather than implementing a gateway at the remote location, the designer chose to implement a directional antenna that would meet the group link budget parameters[182] for the several devices in the remote building. Because the link was not simply between two outdoor antennas, the penetration loss of the building materials had to be accommodated for in the link.

Propagation model tunning and interference testing are the same concepts as used for indoor coverage and capacity planning; however, with an indoor deployment, you are typically concerned about every meter (or close to it) between the gateways and the end devices. This description is not always true in outdoor deployments. For outdoor deployments, the concern is focused on tuning of the propagation models for the target area characteristics (walls, trees, terrain, etc.) and doing this for each target area. The same is true for interference. We are not concerned about some interfering device that has high enough energy emission to impact an area between out gateway and the target area as long as its energy is not sufficiently strong to reach either end of our link.

---

[182] By *group link budget parameters* we mean the maximum required link budget for the least capable device in the remote building. This happened to be the device on the far side of the building from the source gateway and tests showed that if that device had a stable signal for connectivity all other devices in the building would have the same or better signal. This is not always the case as particular sectors of a building may use very different materials that will impact the propagation.

The good news is that long-range protocols are designed to address the issues involved in such communications and they do very well with dealing with interference as long as it does not cover the entire band in which they operate and as long as the interference has a low duty cycle. When these conditions are not true, changing protocols or addressing the interference may become important.

Protocols and Quality Requirements

Quality requirements must make their way into the design process as well. They should be defined during requirements engineering and then the designer must ensure they are implemented in the design. The following quality requirements (non-functional requirements) are often specified:

- **Security:** Security is categorized as a quality or non-functional requirement[183] because it does not provide a direct capability to the user or devices, but it does ensure that the actions taken by the users or devices are performed with privacy, confidentiality, integrity, and availability. Security is discussed independently later in this chapter. It is achieved by designing authentication, authorization, accounting, and encryption into the system and supported by external security systems provided or those provided by the IoT gateways themselves.
- **Performance:** This requirement should be specified in number of messages delivered, percentage of messages delivered, throughput, prioritization of specific messages, and so on. As the designer, you must choose channel plans, Quality of Service (QoS) parameters (where applicable), superframe configuration (where applicable), and other configuration options that will provide for the performance requirements. For example, it may involve selecting a configuration of both the operational parameters and the hardware that can achieve a particular data rate or that can achieve a particular stability level in the link.
- **Reliability:** The probability that a system will work as design is reliability. This is different from availability, which ensures that the system is there. However,

---

[183] While security is a non-functional requirement in relation to a system of interest that is not primarily a security system but rather is intended to be implement in a secure manner, if designing an authentication system, an authorization system, or any other system that is intended to be an actual security system, then implementing a specific authentication or authorization methods becomes a functional requirement. It is all about perspective.

having the wrong system there is not much help. Thorough testing, possibly with proof-of-concept (PoC) or small-scale early deployments, can assist in designing a reliable system that will function as expected. The last thing you want to here is, "Sometimes the system just doesn't give be the right information." This complaint can be avoided by ensuring proper functionality on a consistent basis, which is the ultimate end result of reliability.

- **Durability:** As the designer, your primary focus related to durability is selecting hardware with enclosures that will protect the internal components over a long period in the operating environment. The required device enclosures will be very different in outdoor networks, indoor industrial environments, and office spaces. Another perspective on durability is battery life for battery powered devices. Ensuring quality batteries are used and that the configuration accommodates long battery life is essential. Additionally, using energy harvesting techniques appropriate to the environment (the environment provides the source of energy for sufficient durations each day such as sunlight, wind, etc.).

- **Resiliency:** The ability of the system to continue operations when a component fails is related to its resiliency. Resiliency is typically achieved in wireless IoT design through redundancy. It can also be achieved through system monitoring and performing automatic resets when a device stops responding and so on. Resilience is also a factor of protocol selection. Protocols that perform channel hopping in 2.4 GHz will usually be more resilient than those that use a static channel. Protocols that tolerate weaker signals will be more resilient than those that require stronger signals, and so forth.

- **Expandability/Scalability:** Expandability and scalability are not identical but are both important quality considerations. Expandability is the ability to add devices to the system at a later time that might not be identical to existing devices. Expandability is often achieved because the system supports interoperability. This is most easily achieved by using non-proprietary protocols and standards-based solutions as much as possible. Scalability is the ability to add more of the same devices or different devices to the system and then accommodate them. Adding additional gateways, for example, can be used to scale the system.

While others are also important, these seem to be the most common. For any quality requirements, ensure that you evaluate how to achieve them with your selected solution and design them into the deployment plans.

Synthesis

Synthesis, in systems and product engineering, indicates the combination of various components into a single or unified entity. The goal is to synthesize the requirements, through the framework, into a unified solution for a wireless IoT network design. Figure 6.19 illustrates the synthesis of the wireless network group of the framework concepts and components into a single solution.

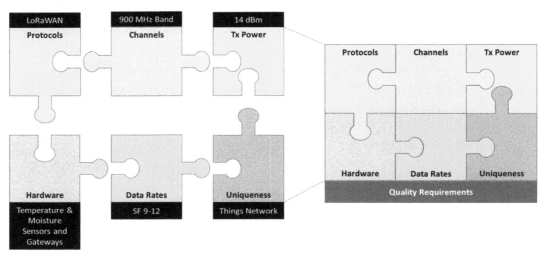

Figure 6.19: Wireless IoT Network Design Synthesis

Synthesis, then, is the final element of the design process for a system. All the design elements should come together to form a functioning system. If they do not, the design element that doesn't fit must be adjusted to allow for proper system operation. For example, you may find that the selected hardware uses an enclosure around the antenna that requires a slightly higher transmit power (if supported) to achieve a stable link. Such adjustments may be made in design or, in the real-world, they may simply have to be made in deployment as they are discovered among the potentially thousands of end devices installed. This is part of the integration process during deployment.

6.4: Using Wireless Design Software

Now that we've addressed the concepts in the design of a wireless IoT network, we can explore wireless design software. Several software solutions exist for the task, and they have varying features and capabilities. The design professional should evaluate available options and choose the one that best fits the need. At CWNP, we do not promote any specific software solution, but will use the iBwave software as an example throughout this section to illustrate the basic concepts. Software applications to consider include but are not limited to:

- iBwave Design
- Infovista Planet RF
- RadioPlanner
- Ranplan Professional
- Forsk Atoll
- Keima Overture
- Siradel S_IoT

A number of additional software packages are available specific to Wi-Fi design should your IoT project use that protocol. For more information on them, see the CWNP CWDP Study and Reference Guide.

The following subsections will focus primarily on indoor design, but outdoor design may be performed as well with terrain maps loaded into software and then implementing the use of proper outdoor technologies. We will explore a simple Zigbee design in this example.

Importing Floor Plans

With most design tools, the process begins by loading the software and creating a new project. The project parameters will include a project name as well as multiple parameters that may constrain the design. Figure 6.20 shows the new project dialog in iBwave Design. You can see that it supports a project name and various details related to project contacts and location. This dialog also allows you to create other parameters such as zones for definition of requirement area type constraints, capacity requirements, units in imperial or metric, and more. Some settings cannot be adjusted until floor plans or maps are imported because they apply to multi-floor or multi-location design scenarios.

With the new project created, the first step commonly taken is to import a floor plan or a map. Design software will often support CAD files that have walls and multiple floors defined. In such cases, the information can be imported with materials assigned. They will also support the import of flat files (JPG, TIFF, PNG, etc.) that require the drawing of walls and objects. We will import a flat file floor plan in this example.

Maps are used more for outdoor network design. Software may support loading maps from online mapping websites that may include topography details (geodata or geographic data). Such information is useful when planning outdoor near-coverage networks using protocols such as 802.15.4, Bluetooth, or Z-Wave or when planning long-range coverage networks using protocols such as LoRaWAN or Sigfox.

Figure 6.20: Creating a New Project

To import a floor plan in iBwave Design, follow this process:

1. With the new project created, a default Floor 1 will exist. Use this to import the floor plan by right-clicking on the tab that reads Floor 1 and selecting properties.
2. You will be presented with the floor plan properties shown in the following image where you can change the name to something more meaningful as we have below.

3. Next, you will import the actual floor plan image by clicking the Browse button next to the Image file field. Select the image containing the floor plan and click Open.
4. Next, click the Config button next to the Scale field so that you can match the scale of the floor plan in the software to reality. You will be presented with a screen similar to the following.

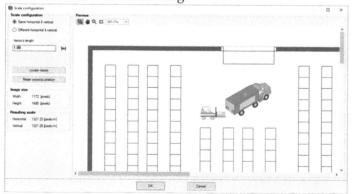

5. Click Locate Vector and then move the endpoints of the vector line so that it spans a distance of which you know the exact distance. Enter that distance in the Vector's length field. Click OK to save the scaling configuration.
6. If you are working with a target facility with ceiling heights above 3 meters (the default), you can change that in the Heights section of the Floor Plan properties dialog.
7. Otherwise, click OK to import the scaled floor plan. You should see results similar to the following.

At this point, you have a floor plan loaded in the software and are ready to begin the design process. Given that we have imported a flat floor plan, we must draw the walls and objects for the facility so that propagation modeling will perform accurately.

Drawing Walls and Objects

In iBwave, wall drawings are performed on the Modeling tab. You may choose to draw objects if the software supports it. Being creative in the tool can help. For example, our floor plan has shelving in a warehouse. If the shelving had solid metal backing for each shelf it would significantly change propagation as opposed to open back shelves. We could draw a metal wall along the backside of each shelf and then set the height to the height of the shelf instead of the default ceiling height. Figure 6.21 shows the propagation modeling results with the "shelf wall" at default ceiling height and Figure 6.22 shows the propagation modeling results with the "shelf wall" at one meter lower than the ceiling. The difference between the two is approximately 14 dB on the other side of the shelving from the Zigbee coordinator. You will need to discover methods for such solutions with the software you choose.

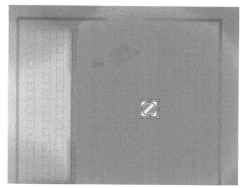

Figure 6.21: Propagation Model with the Shelf Backing at Ceiling Height

Figure 6.22: Propagation Model with the Shelf Backing one Meter Below the Ceiling

To begin drawing walls, select the Modeling tab and perform this process:

1. Click the down arrow under the Draw Walls button.
2. Choose the material for the outside walls. In our case, we'll choose Stone Brick.
3. Click the beginning point for the wall.
4. Move to the ending point for the first wall and click again and repeat.

5. Repeat step 4 until you reach a window or door. When you reach a window or door, press Escape (ESC) to interrupt drawing mode.
6. Move to the next section using the same wall materials and repeat steps 3-4 again.

This same basic process will be used to draw all walls, windows, and doors using the appropriate change in material as you go. After doing this, you will be able to run propagation models that are more accurate because of the materials specification. In Figure 6.23, with the floor plan deselected for viewing, you can now see that we have drawn the primary walls as well as the metal shelf backing for the shelves.

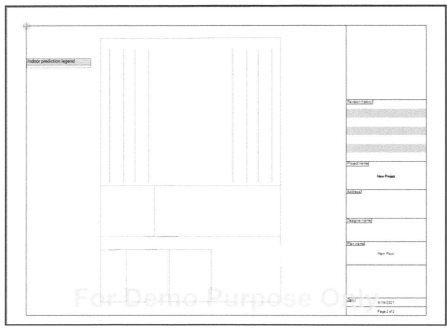

Figure 6.23: Walls Drawn in the Software

Placing and Configuring Gateways

We have the foundation for designing the actual wireless network at this point. We can begin to place signal sources, in our situation, Zigbee coordinators. To place a Zigbee coordinator on the floor plan at the desired location, perform the following process:

1. Expand the Signal Source section in the Parts pane (by default, it's on the left).
2. Choose the Gateway option.
3. In the Select part dialog that appears, select the ZigBee Coordinator in the Model section.
4. In the System section, for Channels, click the ellipses button (…) to set the channel for the coordinator, we will choose channel 11. Then click OK to exit the System Panel dialog and return to the Select part dialog. You Select part dialog should look something like this:

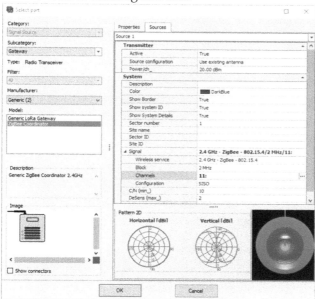

5. Click OK to exit the Select part dialog.
6. Click on the floor plan where you wish to place the coordinator. If you want to place more, click in those locations as well. When you've placed all desired coordinators, press Escape (ESC).

At this point, the Zigbee coordinator(s) have been placed on the floor plan in the desired locations. If you placed multiple coordinators on the floor plan, you could double-click on each one to change the channel settings as required.

305

Showing Cable Runs

Next, for the benefit of the installers, you can place switches on the floor plan and cable runs to those switches. To place a switch:

1. In the Parts panel, expand Network Equipment and choose Switch.
2. In the Select Part dialog, choose the appropriate switch. We will choose a Juniper EX2200-24P-4G switch. Click OK.
3. Click on the floor plan where you wish to place the switch.

Now, you can draw the cable run from the Zigbee coordinator to the switch. To do this:

1. In the Parts panel, expand Cable and choose Twisted Pair.
2. In the Select cable dialog, choose either CAT-5 or CAT-6 and click OK.
3. Hover over the Zigbee coordinator icon on the floor plan and click the blinking blue circle. Move the cursor to the first location where you want the cable to run and click, then the next, and so on, until you reach the switch.
4. When you reach the switch, hover over it and click on a port to indicate that the cable should be attached to that port.

After drawing the cable run, you can click the cable and then use the points to alter its location to your desire if it is not completely accurate.

After creating the cable run, an interesting thing has happened automatically in the background. If you change to the Design Plan table in the project (to the left of what we called the Main Floor), you will see it now shows a switch and the Zigbee coordinator on the design plan. You can drag the objects on the Design Plan view to the locations you desire to represent the architecture. Moving objects here will not impact their locations on the floor plan.

The Design Plan, after some location tweaking, should look similar to that in Figure 6.24 shown on the next page.

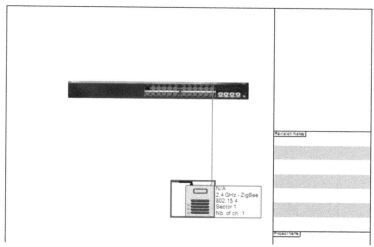

Figure 6.24: The Dynamically Created Design Plan in iBwave Design

Placing End Devices for Simulation

Some tools only provide modeling for gateways and infrastructure devices. Others allow you to place end devices as well. iBwave Design is in the latter category. You can place end devices on the floor plan, which can be useful to model their propagation and ensure that they can reach the coordinator. End devices may not use the same transmit power levels as the coordinator device (which is connected to mains or PoE power) because they may be battery powered and attempting to conserve batter life. This too can be modeled.

You would simply choose the Signal Source section from the Parts panel and then choose End Device instead of Gateway. Configure the output power to match that of your expected end devices (and possibly the antennas) and begin placing them where they will be located on the floor plan.

Figure 6.25 shows the new design so far with Zigbee end devices placed on the floor plan and the Zigbee coordinator placed in a meeting room by the switch. This will be problematic, but we will fix that when we begin to explore the overlays or maps that the software can generate now that we have a basic network design in place.

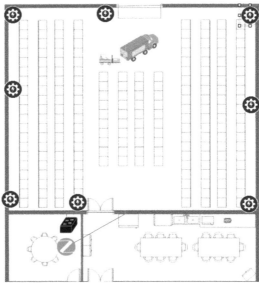

Figure 6.25: Zigbee End Devices Placed on the Design

Generating Overlays or Maps

Everything we've done so far could simply be done in a diagramming tool like Visio or eDrawMax. We are just now entering into the realm of propagation or RF modeling and the real purpose for wireless design software. The propagation or RF modeling will only work with accuracy if the right frequency is being modeled and the building materials have been properly defined. iBwave presents the results of RF modeling on output maps. So, the first step is to create an output map.

To generate output maps, follow this process:

1. On the right panel choose the Prediction tab.
2. Right-click on Output Maps and choose Add Output Map.
3. In the Output Map Configuration dialog, choose the 2,4 GHz - Zigbee – 802.15.4 Wireless Service[184].

---

[184] This is a good point to note something important. If a tool models RF propagation for Zigbee in 2.4 GHz, that means it models propagation for 802.15.4 O-QPSK. Therefore, you can use the same components to model any other RF for 802.15.4-based protocols as well. Just adjust antennas and

4. Accept the default Type of Signal Strength and the default Name of Signal Strength.
5. Change the General Settings if you desire higher resolution settings and a change in Prediction height and click OK.
6. You will now see a Signal Strength output map in the right panel. Right-click it and choose Run.
7. The results will select all Zigbee devices. Expand the output map and choose only Sector 1 (which was the first Zigbee device you place, or the coordinator).

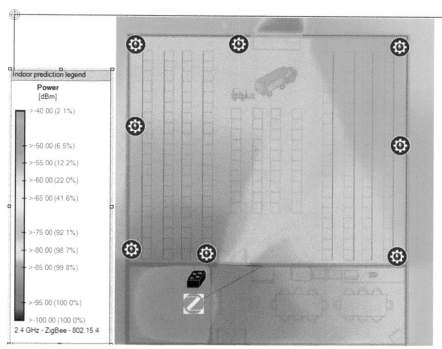

Figure 6.26: Signal Strength Output Map for the Zigbee Coordinator

After performing the steps above you should see results similar to Figure 6.26. We now begin to see the problem with placing the coordinator in the room outside of the

---

output power and you will have the same results. The software is modeling 802.15.4 and not anything that's really constrained to Zigbee.

warehouse. All of the Zigbee end devices are in the warehouse. The coordinator is on the other side of a wall, which happens to be brick in this case as the front office existed before the warehouse was built. In this case, all of the end devices will likely work, but the design is pushing the limits for the Zigbee end device in the upper right corner of the map.

Now, look at the results of moving the Zigbee coordinator into the warehouse instead. The results are shown in Figure 6.27. This latter configuration is clearly much better and we are also wasting less energy where it is not required.

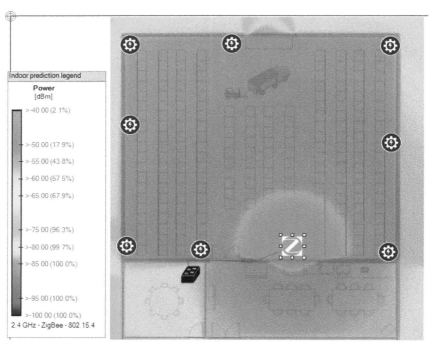

Figure 6.27: The Output Map after Placing the Zigbee Coordinator in a Better Location

Why did we place the end devices on the floor plan? The benefit is that we can now view the maps for those devices as well[185]. Figure 6.28 shows an example of this. The end

---

[185] It is very common for end devices to have lower gain antennas and sometimes significantly lower transmit power than the gateways. This must be tested and accounted for in the design plan. It is another possible motive for using more gateway and lowering the transmit power on them.

device in the upper right corner of the map has a very low transmit power compared to the others modeled after a real-world device in our network. Notice that is signal attenuates very quickly and, in fact, would have likely had problems reaching to coordinator had we not moved it into the warehouse.

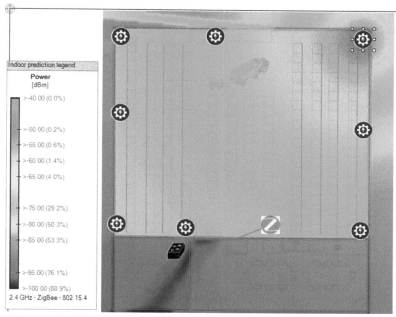

Figure 6.28: Viewing the Output Map of a Specific End Device

Generating Design Reports

With our simple network now designed, we're ready to generate reports. A good design software package should provide for this. Common reports include:

- Equipment Lists
- Output Maps
- Survey Data
- Compliance Reports (these are compliance with requirements, not regulations in most cases)
- Install Reports

The reports available in iBwave Design are on the main Reports tab. The reports will show the output maps, the floor plan with device locations, and other details that you desire. Most software generates a PDF that can be shared with the acquirer and stakeholders to discuss and gain feedback on the design. Figure 6.29 shows the Equipment list report from our basic design.

Equipment List Report

Project name: New Project
Project creation date: 9/19/2021
Design company:
Designer:

Type	Manufacturer	Model	Description	Inventory#	Qty
Cable	Generic	CAT-6	CAT-6 - 24 AWG min. - 100m Maximum Cable Length	N/A	26.55 m
Network Equipment	Juniper	EX2200-24P-4G	24-port 10/100/1000BASE-T Ethernet Switch with four SFP Gigabit Ethernet uplink ports	N/A	1
Radio Transceiver	Generic	ZigBee Coordinator	Generic ZigBee Coordinator 2.4GHz	N/A	1
Radio Transceiver	Generic	ZigBee End Device	Generic ZigBee End-Point Device 2.4GHz	N/A	8

Figure 6.29: Equipment List Report

Additional Tools for Design Processes

Depending on the scenario, additional tools may also be useful. For example, if you must perform a Gateway-on-a-Stick design, the following tools will be quite helpful:

- **Distance measuring tools:** rather than stretching out a tape measure in segments to determine the size of rooms or coverage areas, using a laser distance measuring tool works much better. Wheel-based measuring tools can also be useful. These tools are most important when a proper floor plan is not available, but they are also helpful for validating the floor plan or for acquiring an exact measurement for floor plan scaling in the design software.
- **Cameras:** this is used to photograph important areas or specific machinery that must be recalled during the design. Having the photographs ensures that the item will not be forgotten. We recommend documenting each photo as it is taken. Indicate the purpose of the photo and when it should be used in the design process.

- **Power kits:** this may include batteries, energy harvesting equipment for test purposes, and even generators for remote survey processes. The goal is to have the right equipment to power the design kits during the design process.
- **Personal Protective Equipment (PPE):** this may include safety glasses, safety belts, gloves, protective suits, hardhats, and any other gear required in the facility. Proper PPE should be used at all times to protect yourself and remove liability from the organization.
- **Proof-of-Concept (PoC) kit:** this may include customer devices custom developed for the solution, gateways, coordinators, sensors, actuators, and tags that are intended for use in the production system. They may be used to test functionality during the design process and to ensure that they will work (prove the concept) in the final system.

6.5: Wired Network Design/Planning

This section will not be as detailed as the wireless network design section simply because the wireless IoT designer may not be responsible for all the complex details or may not even be involved. However, the designer should always be prepared to provide the proper individuals or groups with the requirements their design will impose on the wired network that may exist within the organization.

Switching

The most important consideration here is having the switchports available required to connect your IoT gateways. If the gateways are powered via PoE, then having the proper PoE power available is also important. Additionally, some switches support applying QoS markings one Ethernet frames that arrive at the ingress port even if they lack them. This may be beneficial if you require granting QoS priorities to the IoT traffic, but the gateway does not have the capacity to market the frames. You can simply assign a priority to all frames coming in on that interface. Most older switches lack this support, but many newer switches offer it.

Depending on the architecture, the switching structure may be the network backhaul for the IoT network. The *network backhaul* is the portion of the network that provided interconnection to other network segments. We are using this definition because several

IoT protocols refer to the backhaul in this specific way. For example, the backhaul may provide communications between the IoT network and an OT PLC network. Or it may provide communications between the IoT network and the local fog nodes. From a design perspective, the backhaul must accommodate the data throughput that will be generated by the IoT devices.

Routing

With routing, QoS is important as it is at the switching layer. If the incoming packets have priority tags, the router should be configured to carry them over on forwarding egress. This will allow for continued end-to-end QoS. It can be particularly important for reducing round-trip times for IoT devices waiting on a response for actuation.

Additionally, the routers should be able to reach all destinations required of the IoT traffic. This may be a cloud destination or a service location on the network. In all cases, the routers must be configured to allow for it. None of this is unique to IoT, but it is certainly worth documenting in the design plan to ensure that the implementers verify these operations.

IP Management

IP management (IPM) may become an issue if using IPv6-based IoT end devices. The devices may need to participate in a larger pre-existing IPM solution. DHCPv6 may be used for global IPv6 address assignment. It is beyond the scope of this material to provide details of such functions, but it should be addressed in the design documentation if required.

Internet

When cloud-based solutions are used, the Internet will be required for proper operations. Even if public cloud providers are not used, the solution may use the Internet to communicate with remote networks, Internet-based servers, or other systems accessible via the Internet. For example, a mobile health IoT application may access the Internet to get access to current air quality conditions in a city so that this data can be correlated to the health information received about a patient. Then, trends can be detected related to air quality and health conditions for better treatment.

From a design perspective, the most important factor here is that the Internet connection has the available capacity to handle the new workload from the IoT solution. While each device may not generate a large amount of Internet traffic, five thousand devices can quickly consume all the capacity of an insufficient Internet service. As the designer, you can perform the calculations to determine the required capacity and then information the network administration or engineering staff so that they can ensure it is available.

Firewalls

The final component we consider in our framework is the firewall. Firewalls may be used between the IoT network and the rest of the network, between portions of the network, between the internal network and the DMZ, and between the DMZ and the Internet. It is very possible, in some deployments that a communicating starting at an IoT end device and ending in the public cloud could go through three or more firewalls along the journey.

As the wireless IoT network designer, you will need to inform the firewall administrators of the ports and protocols that should be allowed through the firewalls for your solution to work. They need to know, at least, the following two things:

- What ports should be open to and from the Internet?
- What ports should be open to and from the IoT network?

Based on your discoveries when evaluating existing systems, you may be aware of other firewalls that should be opened as well. For example, while planning the database solution for on premises storage, you may have discovered that a firewall exists between the data center and the rest of the network. The firewall administrator may have to allow communications that originate from your IoT gateways through the firewall into the data center.

6.6: Supporting Services Design/Planning

Like the preceding section, this section will not be as detailed, but provides an overview of the considerations related to the supporting services that could be required for a wireless network design.

DHCP

DHCP is only likely to come into significant play in an IoT design in two scenarios:

- DHCPv6 is available and you're using an IPv6 IoT protocol
- You're using Wi-Fi as the IoT protocol

In either of these cases, DHCP can become a very important component in the network. With most IPv6 IoT network solutions, the router or border router can play the role of the DHCPv6 server to the IoT devices and, therefore, you can plan for that directly. However, you must coordinate with the network administrators to ensure you're using routable addresses on their IPv6 network (if one exists).

If you're using Wi-Fi as your IoT protocol, DHCP will become very important. Most Wi-Fi-based IoT devices today still use IPv4 but may support IPv6. In either case, IP is the most likely Network layer protocol you will be using for Wi-Fi-based IoT networks. You should ensure sufficient addresses in the DHCP pools so that the IoT devices can receiver a proper configuration for communications.

DNS

DNS will be important in any cloud-based IoT deployment from the perspective of the gateways at the very least. The gateways must be configured with a DNS server that can resolve the host names of the cloud servers (clusters) so that they can forward IoT data to the cloud. It may play a more important role in 6LoWPAN based networks as well. As the designer, you need only ensure that a DNS server supporting quad-A (AAAA) records is available.

NTP

The Network Time Protocol (NTP) may be used to synchronize clocks among network devices. For many IoT implementations, the gateway will synchronize its clock with an NTP server on the Ethernet network (or the Internet) and the IoT end devices will synchronize their time with the gateway. As the designer, you need to ensure an NTP server is available if required and specify that the gateways be configured accordingly.

SNMP

The Simple Network Management Protocol (SNMP) may be used to manage the gateways on the network. It can receiver alerts from the gateways and in some implementations configure various gateway parameters. You will need to check the vendor documentation to see if SNMP is supported or, if implementing a custom-built gateway, which is still quite common, implement support for SNMP in the operating system you use.

Cloud, Fog, and Edge

Finally, the general concepts of the cloud, fog, and edge should be considered in your Network group of design tasks. This task is about selecting the appropriate location for different processing to occur. From the Network group within the framework, the focus is on ensuring proper configuration so that data can traverse from the IoT end devices to the cloud or the fog or so that edge processing systems are accessible from the end devices or gateways. Let's take a moment to clearly define these terms as they will recur in Chapter 7.

Cloud technologies utilize a combination of virtualization, containers, and virtual networking to provide on demand services for networked applications. Private cloud is used within organizations and public cloud is provided by service providers to organizations. Many public cloud providers have IoT-specific offerings useful for data collection, analysis, and reporting. Many even offer end device provisioning and onboarding services. For these reasons, cloud solutions are appealing to many organizations for their IoT projects.

When it comes to the Network group within the design framework, the cloud solutions most related are device provisioning and onboarding; however, with the decision to use cloud services comes the reality that the cloud connections are now part of your overall network. Certainly, compute is relevant to the IoT solution, but this will be addressed in Chapter 7.

We have the cloud and the edge. The cloud is stuff happening somewhere on a network (either on-premises or on the Internet). The edge is where (or close to where) your devices live (IoT and other devices) and it is roughly equivalent to on-the-edge-of-the-premises. Therefore, there is little difference between edge and private cloud (on-premises cloud) other than the fact that private cloud computing would nearly always take advantage of virtualization or dynamic microservices and edge computing may not.

For example, if you have a physical server processing data from your IoT devices and it is installed on your network, this could be categorized as edge computing. A common concept states that edge computing works on stream data or live data while cloud computing works on big data or stored data. But this differentiation quickly disappears in today's cloud environments where real-time data processing is supported. It is better to define edge computing as on-premises, near the network edge, real-time processing while cloud computing is defined as virtualization-based or microservice-based computing with full abstraction of physical hardware. This is a more meaningful differentiation and applicable to technologies in use today.

The primary motivator for edge computing of real-time data, as opposed to cloud computing of the same, is the reduced latency. If you have IoT actuators in your environment and decisions and actions need to be made and taken within a fraction of a second, sending the decision data to an Internet-based cloud to make the decision may take too much time. Instead, edge computing can be used to provide the decision process with lower round trip times.

That's the cloud and the edge, so what is the fog? Well, imagine that you install a private cloud at the edge of your network. What would you have? Technically, you would have a private cloud; however, you have also implemented a fog architecture. Why differentiate between a private cloud at the edge and a fog architecture? That's a great question and, to be honest with you, I was slow to grasp it. For the first several months of contemplating fog computing and where it fits into distributed application design (that's what it's all about in the end), I was in the "it's just a private cloud at the edge" camp. However, with further thought, and many hours of contemplation, I have decided that it is actually beneficial to think of the fog as something different from a private cloud at the edge. For me, the primary difference is that the fog is integrated with the public cloud (Internet-based cloud) in most implementations, whereas a strictly private cloud is not.

In all this time of contemplation, I spent a lot of time reading what others thought about the fog and it became obvious that many were a bit foggy themselves about what it really is. Consider this definition from the Wikipedia article on Fog Computing:

> Fog computing... is an architecture that uses edge devices to carry out a substantial amount of computation, storage and communication locally and routed over the Internet backbone.

What? So it's just a fancy word for edge computation or edge computing? Well, no. A key word in the definition differentiates it from edge computing: architecture. At first glance, this definition seems to say that fog computing is simple edge computing with a different name but remember that edge computing is just that - computing. Fog computing happens in the fog and if there is a fog, there is more than just computing. Let's explore another definition to understand why this is true.

In a paper from 2011, fog computing was defined as:

> *a highly virtualized platform that provides compute, storage, and networking services between end devices and traditional Cloud Computing Data Centers, typically, but not exclusively located at the edge of network.*
> F. Bonomi, R. Milito, J. Zhu, and S. Addepalli. Fog Computing and Its Role in the Internet of Things. 2012.

Note the use of the term platform in this definition. We have the terms architecture and platform, which begin to reveal how fog is different from "edge" even though it may be implemented at the edge[186]. Edge computing requires only that computing happen at the edge and demands no platform or architecture. A platform is a collection of technologies on which a solution is built. An architecture is the logical design of a system and the connections that exist between the components of a system. You could say that the platform is physical, and the architecture is logical (though the lines here are often blurred as well).

---

[186] Cisco, a primary initiator of the term *fog* for use in these contexts states that *the fog extends the cloud to be closer to the things that produce and act on IoT data* and that *these devices, called fog nodes, can be deployed anywhere with a network connection: on a factory floor, on top of a power pole, alongside a railway track, in a vehicle, or on an oil rig. Any device with computing, storage, and network connectivity can be a fog node.* (Cisco. *Fog Computing and the Internet of Things: Extend the Cloud to Where the Things Are.* 2015) The key element in Cisco's definition is that it's not fog if there is no cloud. In other words, the fog is connected to the cloud and things are connected to the fog. The fog does some of the work, the edge (possibly on the IoT devices) does some of the work, and the cloud does some of the work. The fog nodes connect with both the cloud and the things.

With this understanding, you can see that edge computing is a technology used by the fog architecture or fog platform[187]. In the same way, edge storage may be used by the fog as well as edge networking services. Therefore, fog networking references the networking services of the fog architecture, fog computing references the edge computing of the fog architecture, and fog storage references the storage in the fog architecture. I won't significantly address the previously cited 2012 definition's use of the phrase, "but not exclusively located at the edge of the network," except to say that fog architectures, at least originally, are intended to integrate with the cloud as well regardless of whether they are at the edge or elsewhere on your network. Therefore, you could say that a hybrid cloud (which uses an on-premises private cloud and an Internet-based cloud) implements a fog architecture on premises as the private cloud, given that the hybrid model integrates private and public cloud technologies. In fact, OpenStack, a well-known open-source private cloud solution, has a special committee focused on ensuring it can be implemented effectively for fog solutions.

To further clarify the difference between a fog architecture and a fog platform, consider that you can define a fog architecture without ever stipulating the actual devices, operating systems and specific services that will be used to build it. However, if you implement FogLAMP or OpenVolcano (two open source fog solutions), you are implementing a specific platform.

It is my belief that the phrase *fog computing* should be used literally for *computing in the fog* so that we can have clarity in our discussions. Fogging or "the fog" can be used to reference the architecture or platform that provides fog computing and possibly fog storage and fog networking. It's no wonder that many simply think fog computing is edge computing, because, in most cases, it is nothing more than that (vendors simply jump on the buzzword and call their solution {*today's buzzword*}); however, the fog is something different. It is that solution, inclusive of many of the same components as the cloud, that resides most often at the network edge (specifically the external network edge) and interacts with both the public cloud and the local devices/systems.

---

[187] Edge computing is itself vague however and can be inclusive of IoT end devices themselves, depending on the definition.

6.7: Designing Security

While the specific operations of security capabilities within wireless IoT protocols are covered in detail within the CWICP materials, it is important here for the designer to consider the security in relation to the IoT system under development. Additionally, security is a cross-cutting quality requirement in the system in that it must be implemented everywhere from the physical security of the environment through to the data storage and applications on the network. For this reason, we will focus on security here in its own section in relation to wireless IoT network design and we will address the primary security concerns throughout the entire stack including network access security, hop-by-hop security, and end-to-end security across the wireless IoT solution design framework introduced earlier in this chapter.

Security Overview

Why is security important for IoT? It's a ground-up concept. If security is in place from the start, it's much easier than trying to patch security on top of the system. Security should be integrated into and throughout the IoT solution. Consider the following information.

- Healthcare Service Providers (HSPs) have more breaches than most other common organization types combined. (See Figure 6.30)
- Healthcare is a primary user of IoT.
- The most common identified locations of PHI (Personal Health Information) breaches is email, network servers, paper/films, and electronic medical records (EMR), in that order (See Figure 6.31).
- Three of the top four breach locations are certainly digital and the other of the four (paper/films) is becoming more digital all the time.

These statistics show that digital security, cybersecurity, information security, or whatever you choose to tag it is wildly important for healthcare IoT.

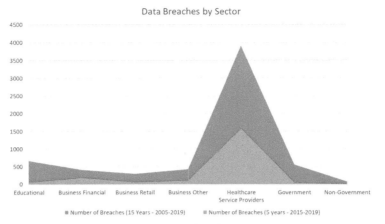

Figure 6.30: Data Breaches in Various Sectors

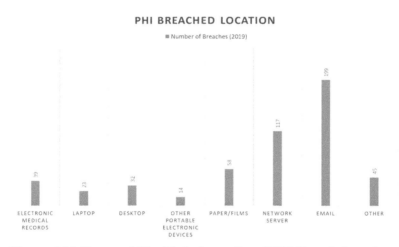

Figure 6.31: Personal Health Information (PHI) Breach Locations

However, while this data[188] shows the important of healthcare IoT (HIoT), other data shows the importance of security in all IoT. Consider the following.

---

[188] Sef, Zerour, Alenezi, et. al. *Healthcare Data Breaches: Insights and Implications*. 2020. Healthcare journal.

- Each device connected to a network is a possible attack point.
- IoT networks often have hundreds or tens of thousands of devices.
- IoT devices are often constrained devices making it challenging or impossible to run client firewalls, anti-malware, and other security services on device[189].
- People, including technology professionals, still tend to underestimate the number of attempted cyberattacks against their systems[190].
- The U.S. Government Accountability Office (GAO) found that 62% of agencies are using IoT technologies and cybersecurity was the challenge most frequently cited by respondents (56.7%)[191].
- Security can cause conflicts among requirements, for example building in more security can result in reduced performance[192].
- IoT is complex (as seen in the multiple levels of information provided in this book and the entire CWNP IoT track) involving protocols at multiple layers, applications, devices, and often Internet technologies.
- Complexity increases the challenge of cybersecurity as more components increase points of ingress and attack surface[193] [194] [195].
- The Mirai malware infected more than 400,000 devices by simply scanning against 61 common hard-coded default usernames and passwords and built a botnet.

Sadly, the preceding list is just the start. We could quite literally fill over one hundred pages with bulleted statistics about security issues related to IoT. This should be enough to help you understand its importance.

Tackling the IoT security issues is easier with a taxonomy of threats related to IoT. Figure 6.32[196] presents such a taxonomy. It starts with the attackers, external and internal, and

---

[189] www.forbes.com/sites/chuckbrooks/2021/02/07/cybersecurity-threats-the-daunting-challenge-of-securing-the-internet-of-things/?sh=7b9277fe5d50

[190] www.techrepublic.com/article/cybersecurity-report-average-household-hit-with-104-threats-each-month/

[191] US GAO. *Internet of Things: Information on Use by Federal Agencies*. 2020.

[192] Voas, Hurlburt. *Third-Party Software's Trust Quagmire*. 2015. Computer journal.

[193] Fortinet. *Network Complexity Creates Inefficiencies While Ratcheting Up Risks*. 2021. Whitepaper.

[194] Baron. *Uncovering the Dangers of Network Security Complexity*. Wired. Half of respondents stated that complex policies ultimately led to a security breach, system outage, or both.

[195] Bralin Technology Solutions. *The Biggest Threat to Cybersecurity is Organization Complexity*. 2017.

then lists common attacks against IoT systems. This list of attacks is a reference list only and not intended to be exhaustive. The attackers initiate attacks (either intentionally or unintentionally) and threaten devices, protocols, data, and software. This diagram is an excellent starting point for introducing effective security to your IoT design. In relation to the diagram, ask the following questions:

- How do I protect my {Devices, Protocols, Data, Software} from {attacker (actor), attack (threat)}?
- What threats are most common in my industry and how do I mitigate them in relation to my {Devices, Protocols, Data, Software}?
- Are my IoT devices constrained or unconstrained or both and how does this impact my security options?
- What known vulnerabilities exist in the chosen protocol(s) and how do I mitigate them?
- How will I secure the data in transit and at rest?
- What maintenance processes should be in place to maintain software security?

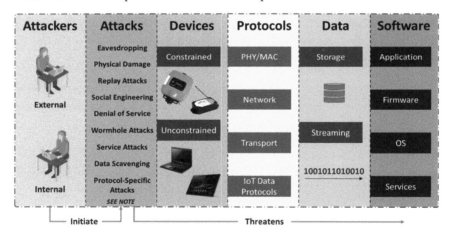

Figure 6.32: IoT Threat Taxonomy

---

[196] Image enhanced from Internet of Things: Architecture, Challenges and Future Directions. *Emerging Trends and Impacts of the Internet of Things in Libraries (pp. 87-104).* Rashid, Gupta, Nazeer, Khanam. 2020. NOTE: The list of attacks/threats is nowhere near exhaustive, but simply illustrative. The designer should develop a list of threats and design security solutions to protect against them.

324

As you can see, a tool like the IoT Threat Taxonomy can be useful in fostering discussions about cybersecurity and as a guide to refocus through the planning process.

Another taxonomy of use is the Cybersecurity Incident Taxonomy, which divides incidents into the source, nature, way of influence, result, and mechanism of action. With this structure, threats and incidents can be categorized and consider based on these groupings. Figure 6.33 represents this taxonomy.

Figure 6.33: Cybersecurity Incident Taxonomy[197]

Now that you understand the landscape in which security must be implemented and the drivers behind the need for effective cybersecurity in IoT, let's consider security for end devices and networking. The IoT Security Foundation has released the Secure Design Best Practice Guides, Release 2, which provides excellent concise guidance on security

---

[197] Modified from Cybersecurity Incident Taxonomy, NIS Cooperation Group, European Commission, 2018 and Evaluation of Comprehensive Taxonomies for Information Technology Threats, SANS Institute, 2018

issues related to IoT design. Some of the information in the following sections is sourced from that document[198].

End Device Security

Physical security is an essential component of end device security in IoT. The IoT devices must be deployed where they can sense the target metric. This demand often requires outdoor installations or installations in common areas within a facility. For this reason, they are susceptible to tampering, theft, and damage. The following guidelines can help to mitigate physical attacks against end devices:

- Remove, disable, make physically inaccessible, or implement proper access controls for any interface of the device used only for administration or commissioning purposes.
- Ensure the device hardware/circuitry is inaccessible to tampering through the use of locked and secured enclosures.
- For IoT devices that travel through the supply chain, consider making them tamper evident.
- Implement environmental monitoring through security controls such as video monitoring, security guards, and employee awareness training.

Using a secure OS is another important factor for end device security. When building your own IoT devices, it's up to you to also select and implement a secure OS. When purchasing IoT devices, known the OS or firmware/software on the device is important. To simplify, you want the minimum OS/software footprint required to perform the function(s) of the device. The following guidelines will help in implementing security for the OS (in this context, OS should simply mean the software that runs the device, whether a true OS, firmware, or software):

- Include only the OS components required to support the device functions.
- Monitor and evaluate security vulnerabilities discovered in the OS and resolve those that apply through patching and updates.
- Use secure configuration by default. Disable everything that is optional within the OS and then enable only those features required. This includes software features, network ports, and local services.

---

[198] IoT Security Foundation. *Secure Design Best Practice Guides, Release 2 November 2019*. 2019.

- Ensure the OS is securely booted.
- Disable write permissions for users and applications to the root file system or system files.
- User accounts and application accounts, if they exist, should have the minimum required permissions.
- Implement an encrypted file system, if available and capable based on device constraints.

Let's discuss secure boot in more detail. *Secure boot* is a solution that uses integrity checks or signature checks against files involved in the boot process and/or used by the operating system during the boot process. Secure boot requires a root-of-trust, which contains the keys used for cryptographic functions and processes. The root-of-trust is usually a hardware security module included in the device that contains the security information used for validation checks.

When vendors provide secure boot, it ensures the boot files, operating system, or firmware is in fact the software from the vendor and has not been modified. When you create your own devices, you will have to implement the entire solution yourself. Generating the keys, creating the signatures, and implementing the checks will be your responsibility. Such a secure boot process should meet the following requirements:

- It cannot be bypassed.
- Verify code before and after it is loaded into RAM.
- Verify the hardware as well as the code to ensure the hardware has not been tampered with.

Credential management is also essential to end-device security. This process starts with the unique identifier of the device itself. Preferably, this unique identifier should be both unique to the device and unchangeable to prevent tampering. The identifier is often used during network joining or onboarding. Additionally, devices may be remotely accessible with usernames and passwords. These credentials should only be used on a secure channel (one that is encrypted) and the password should not be transmitted across the network, if possible (use an authentication protocol that avoids this). They should never be transmitted across the network without encryption. Use strong passwords that are defined as strong in a documented security policy for the security requirements of your organization. If certificates are used in authentication processes, effective certificate management should be implemented with a proper certificate authority trust hierarchy.

Finally, a factory reset function must fully remove all user data and credentials stored on the device with procedures proper for the storage type used.

Update policies should be defined, and secure software updates should be implemented. The policies will define the management of the device throughout its lifecycle and the requirements of update processes. All updates should be performed on encrypted channels if over-the-air updates are used (which is preferred in multi-hundred and certainly multi-thousand device deployments. Update packages should be securely stored and validated before use to ensure they have not been modified. A fail safe mechanism should be part of the update process to ensure that the device is left in an operable state should the update fail.

Given the large number of attacks that occur based on default configurations and default usernames and passwords, it is important to change the default configuration on the devices. This includes any variable that might be used to access the device. The device ID is not likely a variable, but a hard coded setting. If it is variable, it may be changed as well as long as a tracking plan is in place to ensure no duplicates are introduced to the network.

Finally, event logging should be implemented either to a centralized network repository or on each local device so that historic actions can be traced. The logging functions should be a separate process from other device processes and log files should be stored separate from other data files. Storage should be sufficient to store the maximum log history required. Based on storage restrictions, determine the highest priority events to log, such as start-up, shutdown, and login/access attempts. Restrict access rights to the log files so that only special administrative accounts can access them and system services that require them. Passwords and sensitive information should not be stored in log files. Log files should be monitored regularly, preferably automatically, and alerts should be triggered for problems detected.

Additional items may be considered for end device security in specific verticals, but these guidelines will provide a solid foundation for any deployment. You can then add additional security capabilities as required by the specific installation.

Network Access Security

Within IoT, network access security is related to joining the devices to the network. Each protocol defines the joining method for the end devices. With some solutions, devices come pre-provisioned and ready to join the network. This functionality is most common for cloud-based IoT solutions. With other solutions, you will have to provision the devices. For example, you may use a tablet or mobile phone to connect to the device and provision it for access to the network, at which point, the device can join and participate in the network. Provisioning typically includes providing network name or ID information to the device and any credential information the device requires. Joining is the initial connection to the network that establishes a relationship between the device and the gateway or other network component and cryptographic materials for network encryption.

Some IoT devices may have multiple interfaces, both wired and wireless. For example, the IoT device may have an 802.15.4 chipset and a Bluetooth chipset. Or it may have a Wi-Fi chipset and an Ethernet interface. In such cases, all unused interfaces should be disabled within the configuration and in the BIOS if the device has the latter option. Unused but enabled interfaces can be targets for exploit.

All network connections should use encryption within the wireless network. Whether encryption is used on the wired network will be a decision of policy. However, many IoT solutions implement end-to-end encryption at the Application level so that the data is encrypted from the IoT end device to the data processing system, whether that is storage or real-time analytics.

Hop-by-Hop Security

IoT networks often have many more hops between the end devices and the destination of a transmission. Mesh networks may have several hops between the end device and the gateway and then all the traditional network hops still exist between the gateway and the destination. Hop-by-hop security should be considered.

First, all wireless links should be encrypted and most wireless IoT protocols offer this. Some attempt to argue for the lack of need for it due to the dynamic nature of transmissions (such as Sigfox), but too many past experiences have shown that security through obscurity doesn't work, at least not alone.

If encryption is enabled on the IoT network, with practically all IoT protocols, then the wireless hop-by-hop security is addressed. Assuming the protocol allows no devices to participate in the mesh that have not been properly provisioned for the network.

Within the wired network, on a case-by-case basis, you will have to determine whether Layer 2 or 3 encryption is used based on policies and design requirements. However, you can provide security for your IoT transmissions across the wired network, at least for the payload data, by implementing application-level encryption before transmission. Just know that doing this will require more processing power on the IoT end devices.

Firewalls and network segmentation can also be used to provide hop-by-hop security. For example, a firewall could be configured to only allow traffic through the router to which the IoT network is connect from specified sources. For example, a network segment may have servers needed by the IoT network and only that segment is allowed to communicate with the IoT network. In a similar way, access control lists (ACLs) could simply be used in the router to perform a similar control.

End-to-End Security

Security design should not be limited to the Network group in the framework. Rather it should be a fully cross-cutting requirement that reaches end-to-end. While you may not design the specific security solutions for each of the seven design tasks or four design groups, you may need to provide input to them. One of the most severe problems in IoT security today is actually the weaknesses in the IoT Data Protocols design task. Many non-secured MQTT servers have been deployed resulting in tremendous security risk. If your data starts out secure in the network group, but becomes non-secure in the Data group, that is the point where attackers can exploit the solution and easily gain access to the data.

As you are designing the IoT solution, even if you are only planning the Data, Apps, and Operations groups, be sure to provide input and guidance on security solutions that should be considered in those areas. Chapter 7 will provide more information on those three groups of the framework.

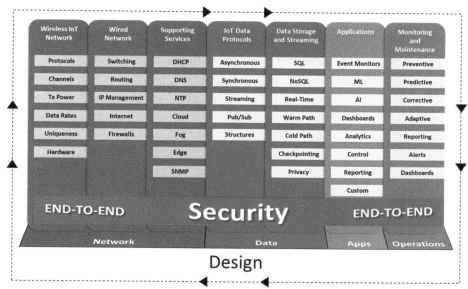

Figure 6.34: End-to-End Security Covers all Four Groups in the Framework

6.8: Integrating with Existing Industrial Networks

Sometimes, you will be required to integrate the wireless IoT solution with existing operations technology networks or industrial networks. These networks have existed for decades, and it is important to understand the terminology related to them. To begin the topic, we will explore the phrase *industrial automation*. The simple and obvious definition of industrial automation is automation within industrial environments. Of course, to fully understand it, more details should be explored. First, what is the meaning of industrial? Second, how is automation implemented in these environments?

The term industry is most accurately used to reference the activities focused on manufacturing products. When we use the phrases "in industry" or "within industry", we typically mean "in factories and other product assembly or creation environments." The products could be final products for sale to consumers or they could be parts used by other manufacturers to create products from those parts. Some organizations operating within the realm of industry are better thought of as assembly operations. They simply take parts manufactured by other organizations and assemble them to create a final product. Whether assembling parts or manufacturing parts and products from raw materials, all such processes can be defined as industrial and function within the business sector of *industry*.

This brings us to the second part of our phrase: automation. To automate something is to cause it to occur without human intervention. A human may start the process, or the process may be started by another process or device, but once started, the process completes without intervention. I could have simply said that automation is "making a process or system operate automatically," but that's just a bit reciprocal. I feel that the clarity of "without human intervention" is a better way to define the term. To be clear, a multi-step process may look more like this:

1. A human does it.
2. Automated.
3. Automated.
4. A human does it.
5. Automated.

In other words, solutions may not be available to fully automate some processes. Enhancements in robotics, artificial intelligence (AI), and human-computer interaction continue to remove the kinds of tasks or operations that cannot be automated. For now, some operations, whether due to safety concerns or expense, continue to be performed manually or with computer assistance by humans.

When considering how computer systems can be used in automation, the model, aptly named "Roles of Computer" in the 1978 paper by Sheridan and Verplank, is a useful place to start. The sketch model provides a way of thinking about automation at different levels from no automation to complete automation.

The indication of the sketch is that a computer can be used to impact the load on humans in several ways:

- Extend - increase the capabilities of a human
- Relieve - lessen the workload of a human
- Backup - backup a human should the human be unavailable or fail to complete the workload
- Replace - completely replace a human in the process

Their model is depicted by the following image (read the entire paper here: https://apps.dtic.mil/dtic/tr/fulltext/u2/a057655.pdf):

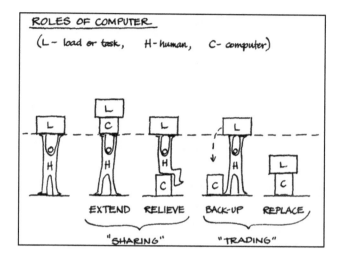

Full automation replaces the human and a computerized system takes on the entire load or task. The other uses of computer systems can be considered as lesser levels of automation or lesser levels of digitization.

Therefore, one can think of industrial automation in different levels ranging from no automation to full automation, passing through extension, relief, and backup along the way. The authors further categorize backup and replace as "trading" as a clarification that these two models trade a computer system for a human. The extend and relieve models are categorized as "sharing" as they implement a computer system alongside the human.

In IIoT, many implementations are related to providing one or more of these levels of automation. For example, an IoT device that resides in a harsh environment to humans could gather data with sensors and pass that information back to a monitoring solution observed by a human. This implementation effectively extends the human's reach into that harsh area. Additionally, when IoT sensors are used through a plant and send their monitoring data back to a central dashboard, it removes the requirement of the human to go to each location and read analog meters to ensure safe operations. This implementation effectively extends and relieves the human of partial workloads.

Control systems have been key to the development of industrial automation. A control system is used to control the output of a machine or another system to meet requirements based on a sequence of events or a specified condition.

Within industrial automation, a control system is used for three primary categories of control[199]:

- **Controlling Variables:** Examples include temperature, pressure, fill level, size of machined objects, weight, and more. The control system can adjust heat levels, flow levels, and other factors to achieve the desired output (temperature, pressure, etc.).
- **Controlling Sequences:** The sequence of events is controlled based on the results or feedback from other events. For example, a conveyor belt may transport a part into a machine and stop until the machine is finished with its processing. If the machine is successful at achieving the desired results, the part is moved out of the machine and to the next phase and the next part is moved into the machine. If a failure occurs in the machine, the part is discarded, or an alert is triggered. The sequence of events is controlled by the control system that monitors the systems and processes. Additionally, the actual sequence of events may be defined by the control system dependent on settings. William Bolton gives the example of a washing machine with the setting to "whites". The machine is activated for a specific sequence based on the control system, which is configured for "whites". If the setting is changed to "towels", a different sequence is initiated by the control system.
- **Controlling Event Occurrence:** In some cases, an event should not occur unless a control system releases it to occur. The example of the washing machine fits here as well. The door lock on the washing machine acts as the control system preventing the wash from starting unless the door is closed. This control systems is particularly important for a front load washing machine.

A complex control system may indeed use all three categories of control or some combination of them.

In the context of industrial automation, a system can be defined as any unit that receives input and produces output. The concern of control is the output of each system, which

---

[199] William Bolton. *Instrumentation and Control Systems, Second Edition.* 2015.

may demand control of the input as well. As an analogy, consider a water filtration system in your home. The input is unfiltered water, and the output is filtered water. You may not understand or care about the actual filtration process that occurs within the system. You are, however, concerned with the output being "better" than the input. The system may use one filter, or it may use several filters. You care only that the output has fewer contaminants in it than the input had originally based on some minimum requirement.

Industrial automation may include systems that machine parts, assemble components, heat elements, cool elements, and more. The control system ensures that these industrial systems are operating such that the output is as desired. The control system may be localized or distributed. A localized control system is like that in a washing machine (considering traditional machines and not those with IoT-enabled remote control). A distributed control system indicates that the parameters of control can be defined from a remote location at the same site or even another site.

Two types of control systems are used within industrial implementations: open-loop and closed-loop. Open-loop control systems receive no feedback from the output, they simply control the input based on a defined value or set of values. Closed-loop control systems do monitor and receive feedback from the output allowing for dynamic adjustments to the input in the achievement of the desired output.

To make is as simple as possible, think of an open-loop control system as one configured to pour 12 ounces of fluid into each container that passes by on a conveyor belt. Think of a closed loop system as one that is configured to pour fluid into a container until it reaches a defined level and the level is monitored during filling. The level is fed back to the input and causes the flow to stop once the level is reached. Open-loop control systems are less expensive and are, therefore, useful in many scenarios when feedback is not required.

Now that you understand the basics concepts of control systems, you can explore two implementations for managing them: programmable (or programmed) logic controller (PLC) and distributed control system (DCS).

A PLC is a computer-based controller that allows for simplified programming of logic, resulting in the control of industrial (or other) machinery. The washing machine example given previously proves useful here. A microprocessor is used in the modern washing machine to enable different motors at different speeds and open different valves at different levels and all at different times depending on the settings configured on the washer. The PLC differs in that it can be reprogrammed by the engineer to perform different functions. For example, one run through a process may be filling bottles that hold 12 ounces of liquid and the next run will be filling bottles that hold 20 ounces of liquid. The program on the PLC can simply be changed to manage these differences.

It is a logic controller because the program effectively says if A then B, or if A and B then D, or if A not B then C, etc. In general terms, the PLC receives input from sensors and provides output to actuators. The internal logic is used to control the actuators based on the program instructions and, possible, input from the sensors. The IEC 61131 standard defines network communication blocks that can be used to standardize the communications within and between PLC systems. These include communications such as:

- CONNECT
- READ
- WRITE
- NOTIFY
- STATUS

A common programming method for PLCs is ladder programming (LAD). LAD uses ladder diagrams to define the operations of the program. It can be implemented as a sort of visual programming where switching circuits are drawn to define the process. While it is beyond the scope of this book to go far beyond terminology definitions, a good resource is Willam Bolton's book, Programmable Logic Controllers, Sixth Edition.

PLC networks may be constructed that allow PLCs to communicate with other computer systems, the Internet, and other PLCs. This option allows for centralized control and management of all PLCs. It is also a key component that allows for integration with wireless IoT systems.

The line between PLCs and DCS solutions have been blurred in recent decades due to the use of PLC networks; however, a fundamental difference is that a PLC is technically a standalone control programmed for a specific task[200] where DCS solutions function as

---

[200] www.automationworld.com/products/control/article/13311313/plc-vs-dcs-which-is-right-for-your-operation

control systems that work through multiple levels to achieve the end result. According to instrumentationtools.com, DCS solutions were designed to control processes and not discrete operations. Another key difference between PLCs and DCSs originally was that PLCs use the ladder logic programming and DCSs use the function block programming. However, that has even changed today as PLCs often support function block programming as well.

The general differences in use between PLCs and DCSs is summarized by Automation World[201] as follows:

- Response Time: PLCs are generally faster and more suitable for real-time safety critical scenarios.
- Scalability: DCSs are generally more scalable, though PLCs can be scaled through PLC networks and can be more complicated than DCSs when scaled.
- Redundancy: DCSs, while not technically more redundant in capability (multiple redundant PLCs can be used), are typically more cost-effective when redundancy is required.
- Complexity of System: More complex systems tend toward the use of DCSs and less complex systems tend toward the use of PLCs.
- High Variability in Processes: PLCs are best used when the process does not change frequently and DCSs are more suited for this.

Supervisory Control and Data Acquisition (SCADA) is a control system that can be used for monitoring and controlling industrial (and other) equipment. One of the key early features of SCADA systems was remote control and monitoring. For example, a sea station could be monitored from land over a satellite connection. SCADA systems are a combination of hardware (PLCs, Human-Machine Interfaces (HMIs), Remote Telemetry Unites (RTUs), sensors, actuators, interfaces, cabling, radios, etc.) and software. Many SCADA software packages are available, but the most popular include AVEVA (formerly Wonderware), Rockwell WinTr, PCVue SCADA, and Siemens SIMADIC.

SCADA systems aggregate data and control related to PLCs and RTUs that are then connected to sensors and actuators (and sometimes more complex aggregated devices). Because SCADA systems aggregate data from multiple sources, decisions related to one

---

[201] Ibid.

component can be made based on the state of other components, which is more challenging with traditional PLCs and RTUs alone (though the line is blurred today).

An entire industrial automation solution may include sensors and actuators reporting to a PLC or RTU. Field HMIs may be available for engineers and staff to monitor the system locally and even provide control input. The PLC or RTU feeds the data back to the SCADA system where central monitoring and control is available and complex logic can be used to make important decisions.

The primary benefit of a SCADA systems is that, typically, PLCs and RTUs are onsite. This results in the need for onsite visits to reprogram them or potentially even to monitor them. While other remote control and monitoring options may be available, the SCADA solutions are most commonly used to gain access to these remote industrial solutions (think, electrical power grid components, oil and gas pipelines, sea stations, or any other remote system that may require control).

Modern SCADA systems integrate with standard enterprise databases, such as SQL Server and Oracle, and even allow for web-based access. APIs may be available for programmatic access into the SCADA system or direct access to data stores may be used to integrate the solution with the enterprise environment.

Ultimately, a good SCADA solution will provide for data acquisition (as the name implies), network data transfer, data representation (for inter-system compatibility), and control. To accomplish this, you implement sensors and PLCs/RTUs, network connections, and SCADA software respectively.

Again, if you've studied for the CWISA, CWIIP, or CWICP exams, you are familiar with the Industrial IoT concept and have likely seen the similarities. An IoT sensor communicates with a gateway/controller/coordinator (like the PLC or RTU) and the data is passed on through the network, often using something like a broker (like MQTT servers), to get the data to the enterprise system (like the SCADA solution). The beauty of IIoT today is that SCADA solutions and modern wireless IIoT solutions can easily coexist. Data from the SCADA system and the wireless IIoT systems can be integrated using common protocols, languages, and databases to provide for a complete view of the systems while taking advantage of the most advanced IoT options available.

The integration options include:

- Shared access to shared databases
- APIs into the SCADA system from the IoT system
- APIs into the IoT system from the SCADA system
- Custom programs and scripts to migrate data

The reality is that many Human-Machine Interface (HMI)/SCADA systems have blurred the line significantly to where they can indeed be categorized as IoT systems themselves[202]. Given that the modern definition of IoT does not necessarily include connectivity to the global Internet and that it requires connecting physical things to a network and gathering information from them or controlling them, SCADA easily fits into the IoT category. The sensors and actuators are the end devices, and the PLCs are the gateways with the SCADA software being the application layer. We can reference it by the old name, to make the incumbents happy, but we can consider it as an existing IoT system with which we must integrate.

Because many have the limited view of IoT as "connected to the Internet," which is in no way a constraint of IoT (and really never has been), they see stark differences between IoT and SCADA systems. For example, one comparison indicates that SCADA uses wired connections to the PLC while IoT uses wireless connections to databases via the Internet. They further indicate that SCADA systems are on premises while IoT systems are cloud-based[203]. However, their contrasts are simply untrue. IoT systems can be wired or wireless and can use the Internet or not. They can be on premises or cloud based as well. As stated previously, a SCADA system actually meets all of the requirements of an IoT system:

- Physical things connected to a network.

---

[202] Marketing materials from SCADA software vendors make it obvious that they feel the need to show that their systems include the capabilities provided by IoT. Non-SCADA IoT solutions pose a threat to their market share and I would not be surprised to see more companies reference SCADA-IoT or SCADA-IIoT to show their integration of traditional SCADA with IoT and avoid replacement of their systems. Indeed, integration seems to be the most logical solution. For more information, see:

https://iiot-world.com/industrial-iot/connected-industry/internet-of-things-and-scada-is-one-going-to-replace-the-other/
https://www.industlabs.com/scada-iiot
https://www.3agsystems.com/blog/iot-vs-scada
[203] https://www.3agsystems.com/blog/iot-vs-scada

- Centralized data monitoring for physical things.
- Control of physical things.

The preceding list describes something. What is it? It is a SCADA system, but it is also a typical IoT system. There is no difference in concept. The difference is in legacy. Just as wireless sensor networks (WSNs) fall into the category of IoT, SCADA systems fall into the category as well when you accept the more common definition of IoT used today[204]. Therefore, it comes down to integration.

As a final note, the same article[205] that compared IoT and SCADA stated that the reason industrial organizations may choose to move to IoT is *the promise offered by the Internet aspect of IoT*. In reality, avoiding outside connectivity has been a primary desire for SCADA systems, in many cases, to increase security and stability of the systems. The article also suggests that IoT is more open and easier to integrate than SCADA systems. However, it's not as simple as that. The reality is that many IoT systems are closed (they use proprietary protocols) and others are open, but they are open using different protocols than other open systems. For example, one IoT system may be engineered based on 6LoWPAN and CoAP, and another may be engineered using Sigfox and the cloud, while yet another is engineered using Zigbee. Integrating such IoT systems is equally complex when compared to integrating different SCADA systems from different vendors.

In the end, it is better to think of these systems as to how the data they generate can be shared, when desired, and how to secure the sharing process rather than attempting to contrast them in such stark manners. If you desire to avoid categorizing SCADA systems as IoT systems (or even WSNs for that matter), the concern of integration remains. It can be achieved through the implementation of data sharing (ETL, databases, applications, etc.) and control conversion (converting control commands from one system to another).

---

[204] I know that some with extensive history in OT will struggle with accepting that SCADA systems meet the qualifications to be called IoT, but this is a semantic issue. SCADA systems definitely have unique qualities that do not exist in all IoT systems, but so do retail IoT solutions, healthcare IoT solutions, and others. Each vertical has different requirements and implementations. Given that SCADA systems are typically wired-only, our primary focus is integrating with those systems and ensuring that they can benefit from the wireless IoT network information and that the wireless IoT end devices can benefit from the SCADA network information.
[205] Ibid.

At times, this will require custom software development, and, at others, integration solutions may be available from the selected vendors.

To summarize, there is no question that the big "I" Internet catapulted IoT into the spotlight, particularly in the consumer space, but many IoT solutions are more related to the little "i" internet or internetwork than they are to the Internet. And remember this when you read about billions of devices being connected to the Internet. These counts do not always include the other billions of devices not connected to the Internet, but that are still devices used to connect things with other things and with management systems. The point? IoT is bigger than the Internet[206].

---

[206] https://www.thingsquare.com/blog/articles/iot-without-internet/

6.9: Chapter Summary

In this chapter, you learned to design a wireless IoT network and the wired network and supporting services based on requirements and constraints. You explored the various components that must be considered in the design process and walked through the basic use of a wireless design tool. Finally, you considered security as an end-to-end essential in wireless IoT design. In the next chapter, you will explore the design or planning of the remaining three groups in the IoT design framework presented in this chapter.

6.10: Review Questions

1. Which one of the following is used in some localization algorithms for wireless devices?
    a. RSSI
    b. Coding
    c. DBPSK
    d. Language translation

2. What is a unique characteristic of LoRa modulation?
    a. Coding
    b. PSK
    c. Chirps
    d. None of these

3. True of False: Wi-Fi offers the highest data rates of common private IoT protocols.
    a. True
    b. False

4. Why is scaling important when importing floor plans into wireless design software?
    a. To ensure that the wall materials have the right attenuation factor
    b. To ensure that the modeling is accurate based on real-world dimensions
    c. To ensure that security is properly designed for the building
    d. None of these

5. True or False: IP management most often becomes an issue because of the use of Bluetooth.
    a. True
    b. False

6. For what is DNS likely to be used when implementing cloud-based IoT?
    a. Assigning IP addresses
    b. Assigning gateways
    c. Resolving host names
    d. Synchronizing clocks

7. Why is it important to define building materials in your wireless design software?
    a. So the customer will know the cost of erecting the building
    b. So the software can properly model RF propagation
    c. So the software can recommend the best vendor gateways
    d. So the software can inform you of walls that must be removed

8. True or False: End device security should include physical security considerations.
    a. True
    b. False

9. What is the term used for the colorful overlay used in design software to show various metrics related to the wireless network?
    a. Heatmaps
    b. Floor plans
    c. Peer-maps
    d. Reports

10. What is not typically included in project parameters within wireless design software?
    a. Gateway locations
    b. Project name
    c. Measurement units
    d. Project location

6.11: Review Answers

1. The correct answer is A. RSSI is used by many localization algorithms.
2. The correct answer is C. LoRa is the only protocol covered in the CWNP IoT track that uses chirps in the modulation.
3. The correct answer is A. The statement is True.
4. The correct answer is B. Without scaling the software cannot properly perform RF modeling.
5. The correct answer is B. The statement is false.
6. The correct answer is C. DNS is a name resolution service and protocol.
7. The correct answer is B. If the building materials are not defined, the software is unaware of the attenuation and reflection factors of that material.
8. The correct answer is A. The statement is true.
9. The correct answer is A. These overlays are called heatmaps and often include signal strength, SNR, expected data rates, throughput rates, number of APs, CCI, and more.
10. The correct answer is A. Gateway locations are defined on the floor plans and not in the project parameters.

Chapter 7: Designing the IoT Solution Beyond the Network

Objectives Covered:

3.5.1 Required infrastructure hardware and software

3.5.3 Recommend robust security solutions

Throughout this chapter you will notice the use of the words plan, plans, and planning. This usage is not accidental. While you may have the skills to design the wired network, the end devices, the supporting services (enabling systems), and more, the focus of CWIDP is on planning these components more than designing them. The difference is in the role you will have as a wireless IoT designer. You must ensure that you can perform the design of the wireless network supporting the IoT end devices. However, you may not be the designer for every other component. The design of database solutions, integration through IoT Data protocols, and other supporting services is discussed more in CWIIP.

With that clarification, this chapter still provides important details for these various components. They are covered with enough information to allow you to plan for them from a recommendation-based perspective in your design plans. For example, you can suggest the use of MQTT with particular QoS parameters to meet the requirements of the IoT solution under design.

The chapter covers a wide range of important topics including IoT Data protocols, data storage and streaming, applications, and monitoring and maintenance. All of which are essential to an end-to-end IoT solution. We will begin by discussing IoT Data protocol planning.

7.1: IoT Data Protocol Planning

IoT Data protocols have been referenced several times throughout this book already. They were first introduced in Chapter 1 definitionally, and they have been referenced in relation to existing systems, requirements, and end-to-end security. In this section, we will provide sufficient knowledge of how these protocols work for design planning. IoT Data protocols are often called *middleware* protocols because they often act as a bridge between applications. Instead of talking directly to each other as peers, they may converse through the IoT Data protocol. This structure can be conceptualized as an application layer star or hub-and-spoke network for many IoT Data protocols. One exception is the DDS protocol, which behaves more like an application layer mesh (more on that later). Additionally, some IoT Data protocols (when used for IoT data) are indeed peer-to-peer protocols. Examples include CoAP and HTTPS, where requests and responses are sent directly between peers.

The protocols covered include:

- MQTT
- AMQP
- DDS
- CoAP
- HTTP/HTTPS

Many IoT Data protocols use a publish and subscribe (pub/sub) model. Between the publisher and subscriber is a message notification and delivery service, often called a broker or simply a server. For example, a publisher publishes message pm_1 to the server for topic t_1 to which a subscriber has subscribed. Message pm_1 is sent to the subscriber asynchronously, meaning that it does not have to be delivered immediately but when the subscriber is available to receive it. Figure 7.1 represents this communication model.

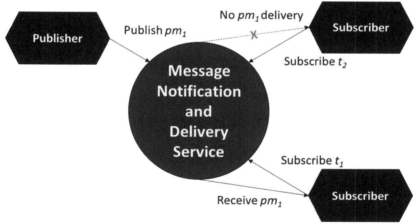

Figure 7.1: Basic Pub/Sub Model

Message Queuing Telemetry Transport (MQTT)

MQTT is a publish-subscribe protocol. Within MQTT, the server is sometimes referred to as the broker. As previously listed, there are three main components to the publish-subscribe model:

- Publisher
- Subscriber
- Server (or Broker)

In most deployments of MQTT it uses TCP/IP as its transport layer. Clients establish TCP/IP connections with the server or broker, but they do not establish direct TCP/IP connections with each other.

Clients or devices can publish or subscribe to topics to transmit or receive messages. *Subscriber* and *publisher* are special roles of a client. Clients participate in a sequence of steps to fulfil their roles. They open network connections to the server and transmit a message to the server. This act results in published messages that other clients might be interested in. Clients also subscribe to messages they might be interested in receiving. Clients can also send unsubscribe messages to the server indicating they do not want to receive messages from topics they are no longer interested in. It is important to note the publisher and subscriber are not required to be connected to the infrastructure at the same time. In addition, clients do not have direct connections with each other, but only direct connections with the server.

A server is a software application running on a computer. You may find the MQTT server referred to as broker in other documentation or articles. This is simply a matter of semantics and a reference to older standard terminology. As version 5.0 of the MQTT standard refers to this device as server, we will do the same throughout the rest of this chapter. Computers could be local devices running Windows, macOS, or Linux. MQTT is also often run in local virtual machines. The application could run on low-cost hardware like a RaspberryPI or NanoPI Neo2. It could be run on-premises or run in a cloud-based service such as Amazon Web Services. Examples of some common MQTT server software are listed below:

- Mosquito (https://mosquitto.org/)
- HiveMQ (https://www.hivemq.com/)
- ActiveMQ (http://activemq.apache.org/)
- AWS IoT (https://aws.amazon.com/iot/)
- CloudMQTT (https://www.cloudmqtt.com/)
- Node-Red (https://nodered.org/)
- Adafruit IO (https://io.adafruit.com/)

In a pub/sub model, clients are not aware of each other and do not know the IP address or internal network details of other clients. This detailed information is only exchanged between clients and the server. The pub/sub model supports one-to-one, one-to-many, and many-to-one. Publishers need only send one copy of a message to the server and the server takes care of replicating copies of the message to all subscribers. When the server receives a published message, it applies a filter. Based on the filter rules, the server determines which messages need to be published to all subscribed clients.

Servers act as an intermediary or relay between clients that publish application messages to clients that have made subscriptions. A server's role begins by accepting network connections from clients. Servers accept application messages published by clients. Servers process the subscribe and unsubscribe message requests from clients. Using filters, servers forward or publish application messages to any clients that subscribes to matching topics. Finally, servers can close network connections from the clients.

There are several benefits to using a centralized server. For one, servers can be used to eliminate vulnerable and insecure client connections by making it necessary to use secure connections combined with authentication using username/password or certificates. Servers allow the infrastructure to easily scale from a single client to 1000s of devices without the need to touch every device each time a new device is added. This can all be done while placing minimal strain on the network for even the most tenuous networks such as a satellite network.

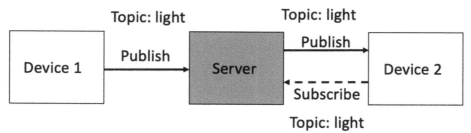

Figure 7.1: The MQTT Server

MQTT *topics* can be thought of like a channel - they are the glue that connects the publisher and subscriber together and are used to express interest for incoming messages. The topic namespace is an unmanaged way to identify messages. It is referred to as unmanaged because there is no enforced format or standard to follow. Each client determines the topic or topics of that device's messages.

While allowing a high level of flexibility, an unmanaged namespace requires programmers and integrators to be disciplined with consistent and specific topic names. That is, the applications are in charge of the topics, and it can get out-of-hand quite quickly with multiple topics containing the same data and bring confusion to the overall data structures and integration possibilities. Topics are created by publishers and subscribers when a client publishes a message or when a client subscribes to a topic, but they are not permanent. Topics need to be shared with other devices that interact with the server since those devices will need to know what topic(s) to subscribe to and what to publish. Topics can be dynamically created when a client subscribes to a topic or when a client publishes a message to a topic with the retain message flag set to True.

Topic names are case sensitive strings written in UTF-8 format. For example, the topic 'house/HVAC' is not equal to 'house/hvac'. MQTT topics are structured in a hierarchical format using the slash ("/") as a delimiter to mark different levels within the topic. Spaces are valid characters that can be used within the topic name; however, this is not a best practice. Wildcards can be used to filter the topics of interest or request multiple topics at the same time.

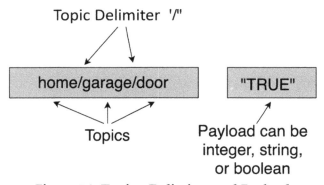

Figure 7.2: Topics, Delimiter, and Payload

There are two types of wildcards and an exception character.

'+' (plus character) – single level wildcard

'#' (octothorpe character) – multi level wildcard

'$' (dollar sign character) – reserved for internal statistics of the MQTT server

A client may subscribe to individual or multiple topics. To better explain how wildcard characters work, we will use some examples with a list of the following individual topics:

```
house
house/HVAC
house/room1/door
house/room1/window
house/room2/door
house/room2/window
house/garage/door
house/garage/light
```

Subscribing to 'house/+', would return the following topics:

```
house/HVAC
house/room1
house/room2
house/garage
```

Subscribing to 'house/room2/+', would return the following topics:

```
house/room2/door
house/room2/window
```

Subscribing to 'house/+/door', would return the following topics:

```
house/room1/door
house/room2/door
house/garage/door
```

Subscribing to '#', would return the following topics:

```
house
house/HVAC
house/room1/door
house/room1/window
house/room2/door
house/room2/window
house/garage/door
house/garage/light
```

Subscribing to 'house/#', would return the following topics:

```
house/HVAC
house/room1/door
house/room1/window
house/room2/door
house/room2/window
house/garage/door
house/garage/light
```

It may be easier to think of the topic namespace as a topic tree with increasing levels of depth to represent more specific devices, locations, or configurations. An example of a topic tree for the topics listed above is shown in Figure 7.3.

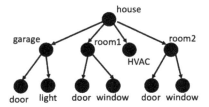

Figure 7.3: Topic Tree

The '$' character is reserved to signal topics with a unique purpose. Any topic beginning with the '$' character is not included in the multilevel wildcard ('#') match, rather the $-

symbol topics are reserved for internal statistics of the MQTT server. In addition, clients are not permitted to publish messages to these reserved topics. There is no official standard for using $-symbol topics, but the MQTT Github wiki website[207] shares some examples using $SYS:

- **$SYS/broker/load/bytes/received**: The total number of bytes received since the broker started.
- **$SYS/broker/load/bytes/sent**: The total number of bytes sent since the broker started.
- **$SYS/broker/clients/connected**: The number of currently connected clients.
- **$SYS/broker/clients/disconnected**: The total number of persistent clients (with clean session disabled) that are registered at the broker but are currently disconnected.
- **$SYS/broker/uptime**: The amount of time in seconds the broker has been online.
- **$SYS/broker/version**: The version of the broker. Static.

While programmers and system integrators have the flexibility to create topics using practically any name, there are some suggested best practices to follow sourced from HiveMQ.com:

1. Never use a leading forward slash.
2. Avoid using spaces in a topic (just because you can, does NOT mean you should).
3. Keep topics short and concise.
4. Use only ASCII characters and avoid using non-printable characters.
5. Embed a unique identifier or the ClientID into the topic.
6. Do NOT subscribe to '#'; this is like "debug all".
7. Do NOT forget about extensibility.
8. Use specific topics, NOT general ones.

---

[207] https://github.com/mqtt/mqtt.github.io/wiki/SYS-Topics

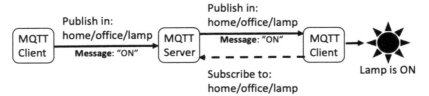

Figure 7.4: Sample MQTT Message Exchange

A handful of different messages can be exchanged between clients and servers. Some messages contain payloads and others are simply used to send control leveraging flags within the packet, absent any payload. Payloads are the content of messages and are data-agnostic, meaning that the payload could be in the format of: binary, integer, string, JSON, XML, or image. For enhanced security purposes, the payload could also be in an encrypted format.

When a client publishes a message on a topic to the server, the server will, in turn, relay the message to all connected clients that have subscribed to that topic. Once the message has been sent to those clients, it is removed from the server. If no clients have subscribed to the topic or those clients are not currently connected, the message is also removed from the server as the server does not store messages. However there are some exceptions: a retain flag can be set on the message, persistent connections can be enabled, and selected QoS levels can result in the messages being temporarily stored by the server. We will discuss more details on those exceptions later in this chapter.

An example of typical message call flow is shown in Figure 2.14.

Now, let's discuss the MQTT control packets. The minimum control packet size is only 2 bytes with a single byte control field and a single byte packet length field. There are four main components of the MQTT Control Packet:

1. Fixed Header – present in all MQTT Control Packets
2. Packet Length – present in all MQTT Control Packets
3. Variable Header – present in some MQTT Control Packets
4. Payload – present in some MQTT Control Packets

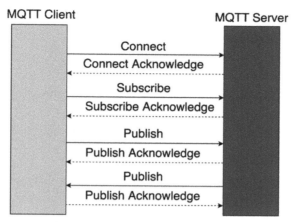

Figure 7.5: MQTT Flow Example

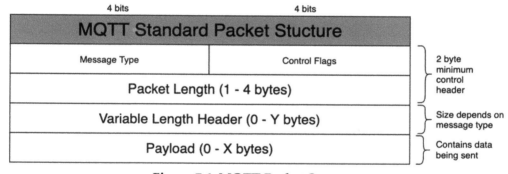

Figure 7.6: MQTT Packet Structure

There are 16 different control packet types defined within MQTT though the first type is reserved. The 15 available control packet types are:

- CONNECT: Connect to a server
- CONNACK: Acknowledge connection
- PUBLISH: Publish to server (client-to-server)
- PUBACK: Acknowledge publication (server-to-client)
- PUBREC: Publication received (server-to-client)
- PUBREL: Publication complete (client-to-server)
- PUBCOMP: Publication complete (server-to-client)

- SUBSCRIBE: Subscribe to a topic
- SUBACK: Acknowledge subscription
- UNSUBSCRIBE: Unsubscribe from a topic
- UNSUBACK: Acknowledge unsubscription
- PINGREQ: Notification of life, check for life, keep connection alive
- PINGRESP: Response to PINGREQ
- DISCONNECT: End connection (clean client request for disconnection)
- AUTH: Authentication exchange

MQTT traffic is generally classified and marked as best effort (BE) at the TCP layer. Even though TCP/IP is designed to provide guaranteed delivery, data loss can still occur on links with intermittent network connectivity where TCP connections break and messages are lost in transit. To deal with the possibility of intermittent packet loss, MQTT includes three possible QoS levels to differentiate the kind of guarantee a message has for reaching the intended recipient.

The first is QoS Level 0. It is officially called as 'at-most-once' delivery. This level is also called 'fire and forget'. Level 0 is intended for use on very stable networks or scenarios when it is acceptable to lose occasional messages. Messages are not retained by the server when level 0 is set. Think of level 0 like requesting your kids clean their room, but never receiving a confirmation that they received the request.

QoS Level 1 is called 'at-least-once' delivery. Messages are guaranteed to be delivered, but the client may publish duplicate messages if it does not receive a PUBACK (publish acknowledgement) frame. Level 1 is intended for use cases where applications have error handling capabilities to deal with receiving duplicate frames or when the applications need to guarantee delivery of messages without the burden of additional frames found in level 2. An example of a level 1 application is alarming functions such as publishing whether a door is open or closed. Think of level 1 like repeatedly requesting your kids clean their room until you receive an acknowledgement that they heard the request.

QoS Level 2 is called 'exactly-once' delivery and is the highest level of delivery, but it comes at the expense of generating the most overhead. Level 2 is intended for applications that need to receive messages exactly once and cannot deal with receiving duplicate messages. Level 2 is a combination of level 0 and 1. Level 2 messages can be stored by the server, but storage requires additional message overhead to inform the server when a message can be deleted. Instead of sending a PUBACK, with level 2, a

PUBREC (publish received) message is returned. Furthermore, the second step of the level 2 protocol exchange occurs when the client is ready for a published message to be deleted. A client will send a PUBREL (publish release) message which is followed by a PUBCOMP (publish complete) message response from the server. Think of this as a civilized conversation between a parent and child where the child acknowledges receipt of the room cleaning request. This exchange is followed up with 'thank you' and 'you're welcome' acknowledgments[208].

MQTT has some security features included as part of the protocol. The two default ports used by MQTT (TCP/IP) are port 1883 for unsecured connections and port 8883 for secure connections. The secure port uses SSL/TLS to encrypt data sent end-to-end between the client and the server, much like HTTPS. In addition, clients can be forced to authenticate with the server by including a username and password value. Optional support is available for the use of certificates. Enabling certificates requires clients to be provided with a signed certificate that matches that of the server.

Most server software supports a redundant server that can provide automatic redundant backup functions. Server applications can also be set up to load balance connected clients to redundant servers on-premises, in the cloud, or a combination of both. This can include the ability to shift client load as servers get overwhelmed or go offline. All of this makes the entire model very scalable and flexible to change and growth. In a lab environment or during a proof-of-concept exercise, a single server may be sufficient. When integrating a production deployment, it is best practice to deploy servers with redundant capabilities to mitigate any unexpected downtime.

Of course, it must be remembered that application code is ultimately behind all the MQTT communications. The design of the application can include resiliency. For example, the application on the publisher side may have a list of MQTT servers to which it may publish, prioritized from highest to lowest. This list could include two servers or two hundred servers. It doesn't matter because it's the application code that handles it. On the subscriber side, the same is true. If a PINGREQ message is not acknowledged by

---

[208] Notice the dependency on application logic in these QoS levels. While the QoS levels are set in the messages and specific message types are used between the publisher and the server, the subscribers must have code that handles the proper processing. For example, knowing that a particular message is already received and not taking processing actions related to an already received message. From topics to message processing, MQTT leaves much in the hands of the applications that access the published messages.

the server, the subscriber can quickly form a connection with the next server in the list and subscribe to the same topic to continue receiving messages. MQTT is quite flexible as it is ultimately an unintelligent man-in-the-middle that is intentionally created to achieve such flexibility.

MQTT-SN

MQTT-SN is the other version of MQTT and is designed to work over UDP, ZigBee, and other transport protocols. The latest release of MQTT-SN was version 1.2[209], published in 2013 although it has not been significantly developed since. 'SN' stands for 'sensor networks' as MQTT-SN was intended as an evolution of MQTT adapted for peculiarities commonly experienced when integrating wireless sensor networks (WSN). Peculiarities include low bandwidth, high link failures, short message length, low-cost, battery-operated devices, or devices with limited resources such as CPU and memory.

As stated in the MQTT-SN v1.2 standard, differences with MQTT are characterized as[210]:

1. The CONNECT message is split into three messages. The two additional messages are optional and are used to transfer the Will topic and the Will message to the server. This topic/message combination is published to each subscriber when they first connect to the server and subscribe to the topic.
2. To cope with the short message length and the limited transmission bandwidth in wireless networks, the topic name in the PUBLISH messages is replaced by a short, two-byte long "topic ID". A registration procedure is defined to allow clients to register their topic names with the server/gateway and obtain the corresponding topic IDs. It is also used in the opposite direction to inform the client about the topic name and the corresponding topic ID that will be included in a following PUBLISH message.
3. "Pre-defined" topic IDs and "short" topic names are introduced, for which no registration is required. Pre-defined topic IDs are also a two-byte long replacement of the topic name; their mapping to the topic names is known in advance by both the client's applications and the gateway/server. Therefore, both

---

[209] http://mqtt.org/new/wp-content/uploads/2009/06/MQTT-SN_spec_v1.2.pdf

[210] Ibid.

sides can start using pre-defined topic IDs; there is no need for registration as in the case of "normal" topic IDs mentioned above. Short topic names are topic names that have a fixed length of two octets. They are short enough for being carried together with the data within PUBLISH messages. As for pre-defined topic IDs, there is also no need for registration for short topic names.

4. A discovery procedure helps clients without a pre-configured server/gateway address to discover the actual network address of an operating server/gateway. Multiple gateways may be present at the same time within a single wireless network and can cooperate in a load-sharing or standby mode.

5. The semantic of a "clean session" is extended to the Will feature. For example, not only client are subscriptions persistent, but also Will topics and Will messages. A client can also modify its Will topic and Will message during a session.

6. A new offline keep-alive procedure is defined for the support of *sleeping* clients. With this procedure, battery-operated devices can go to a sleeping state during which all messages destined to them are buffered at the server/gateway and delivered to them when they wake up.

There are three components that comprise the MQTT-SN architecture:

- MQTT-SN clients
- MQTT-SN gateways (GW)
- MQTT-SN forwarders

Like MQTT, MQTT-SN clients do not connect directly with each other. However, MQTT-SN clients connect to an MQTT server via an MQTT-SN GW using the MQTT-SN protocol. An MQTT-SN gateway acts as a translator between MQTT and MQTT-SN. If the MQTT-SN gateway is not directly attached to the network, clients can connect to a forwarder which simply encapsulates the MQTT-SN packets and forwards them unchanged to the gateway.

MQTT connections are reserved end-to-end between MQTT-SN clients and the MQTT server. In the transparent model, each MQTT-SN client establishes an individual session with the MQTT server. Due to the end-to-end nature of these sessions, all functions and features implemented by the server can be offered to the client. While the transparent gateway is arguably simpler to implement, compared to the aggregating gateway, it does place the burden on the MQTT server to support unique sessions for each active client.

This may present a design challenge with some MQTT servers that limit the number of concurrent active sessions.

Instead of establishing individual sessions for each active client, an aggregating gateway terminates all client sessions and establishes one single session to the MQTT server. Aggregating gateways can decide what information to pass along to the server, thus filtering some fields from reaching the MQTT server. Arguably a more complex gateway implementation, compared to transparent gateways, aggregating gateways do offer the benefit of reducing the number of concurrent active sessions that need to be maintained on the MQTT server.

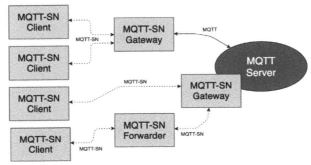

Figure 7.7: MWTT-SN Architecture

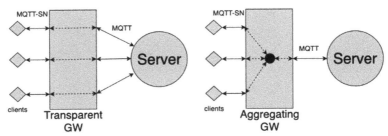

Figure 7.8: Transparent Aggregating Gateways

MQTT also supports operation over a WebSocket with some strict guidelines. The network connection must be closed if the MQTT control packets are not sent in a WebSocket binary data packet. Binary data packets are the only format supported. Individual MQTT control packets can span across multiple WebSocket data packets and receivers must support packet aggregation to reassemble MQTT control packets at the destination. Client and server must both support the WebSocket subprotocol MQTT and each peer must refer to it as "mqtt".

Advanced Message Queue Protocol (AMQP)

AMQP (amqp.org) is a publish-subscribe, open standard, message-oriented protocol. AMQP stands for Advanced Message Queue Protocol with the most recent version 1.0 approved as a standard by both OASIS and ISO (ISO/IEC 19464:2014).

AMQP is like MQTT in several aspects. Both protocols are asynchronous based on TCP and may optionally use TLS for security at the transport layer. They are also supported by a wide range of platforms and programming languages. While AMQP does not provide a hierarchical structure, it remains true to its name and provides message queuing as a method to buffer or store messages. MQTT messages are published with a single global topic or namespace, whereas AMQP messages may be forwarded to several different queues. AMQP is considered a binary protocol – meaning that everything sent over AMQP is binary.

As with MQTT, there are three main components to AMQP: producer, consumer, and broker. Messages originate from a client, called the producer, who transfers message(s) to an exchange. The exchange and queue buffers exist on the broker who receives message(s) from the producer. Using rules defined by the exchange and routing keys provided by the producer, messages are forwarded to one or more subscribers for consumption. Queues provide the benefit of buffer storage as messages are waiting to be consumed. Queues may be created with many different attributes including durable, auto-delete, and exclusive. Various queue markings define how or if messages are to be stored if subscribers are not active or the broker restarts.

AMQP uses a type system for data communicated within the messages. The type system includes primitives for the most commonly used data types, such as null, boolean, integers of various lengths, decimal values of various precisions, strings or characters, times, binary data, lists, and more. Additionally, the protocol supports type descriptors so that the primitive types may be extended by constraint to form a custom type. For example, you may wish to use a descriptor with the string primitive to define a custom type that always contains a Web URL.

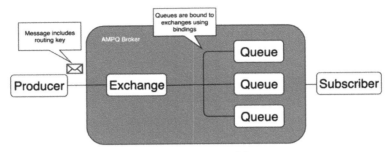

Figure 7.9: AMQP Architecture

AMQP is supported by Azure IoT Hub, AWS Amazon MQ, and the CloudAMQP dedicated cloud service, among others. RabbitMQ is the most popular open-source solution and, interestingly, most cloud providers use it as the foundation for their offerings.

If you would like to get more hands-on experience with AMQP, RabbitMQ (rabbitmq.com) is one of the most widely deployed open-source message brokers used with AMQP.

Data Distribution Service (DDS)

Data Distributed Server (DDS), managed by the Object Management Group (OMG)[211] who host the DDS Foundation, uses a data-centric publish-subscribe (DCPS) model to exchange messages between nodes[212]. It serves as middleware or an API standard for machine-to-machine communication. However, DDS is more data-centric than other IoT Data protocols and goes beyond typical publish-subscribe models to attach contextual based information to messages.

Contextual information allows a filter to be applied based on content, age, and/or lifecycle. Source filtering can include content-based and time-based filters. Filtering can significantly reduce communication overhead in constrained systems. For example, if you are using DDS for a temperature sensing solution and the sensors publish to readings, you can filter to indicate that you only desire to receive publication if the

---

[211] For context, the OMG organization is the same organization that brought us CORBA and UML. They are a well-established organization dating back to 1989.
[212] DDS Specification 1.4

temperature is outside of a specified range. This configuration would limit the received publications to those that may indicate a problem.

As part of managing resources, DDS supports a flexible QoS mechanism which can dynamically respond to resource changes and limit the distribution of messages to all subscribed nodes, instead of sending messages to only those nodes that really need the messages. Changes to message distribution are communicated to all participants. DDS uses topics to capture the properties of data as domain-wide information defined within a <name, type, qos> tuple.

DDS applications work collectively by autonomously and asynchronously reading and writing data on a data space, known as the Global Data Space (GDS), as shown in Figure 7.10. The representation in Figure 7.10 is conceptual. There is no service broker or server in the middle. Instead, DDS uses peer-to-peer communications forming a sort of mesh of data.

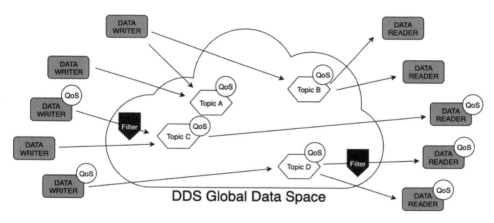

Figure 7.10: DDS Global Data Space

Topics in DDS are different from topics in other pub/sub models as they are in the form of a triad: a type, a unique name, and QoS policies. They type parameter effectively defines the data type or structure. The unique name is the name used by subscribers to subscribe to the topic. The QoS policies can be quite complex, but the system will use defaults defined in the specification if they are not explicitly defined in the publication. As an example of the potential complexity, DDS supports more than 20 QoS policies that may be used together or individually. To achieve particular QoS levels of intent, you must use the right combination with a published topic. For example, to guarantee that all publications reach all subscribers at some point, you must implement the Reliability policy with the Reliable variant (instead of the Best Effort variant) and you must implement the History policy with the Keep All variant (instead of the Keep Last variant). If you further want to impose a priority of delivery on the publication, the QoS policies for Transport Priority, Latency Budget, and Deadline are typically used.

The goal, here, is not to explain everything about DDS, but to provide sufficient information for you to understand the reality of its implementation. If you require advanced control in your IoT Data protocol, DDS is likely the very best option. However, with that advanced control comes significantly greater complexity than the other protocols discussed here and you should be prepared to accommodate more time in the deployment process based on this fact.

Additionally, instead of a single grouping of all publications, DDS supports domains and partitions. Therefore, the data space can be divided to that they can participate only in a partition. You can think of the partition like a subject for which you wish to gather information.

More information about DDS can be found at www.dds-foundation.org.

Constrained Application Protocol (CoAP)

The Constrained Application Protocol (CoAP) is also a popular IoT Data protocol. Based on a Client/Server model it is intended for use with constrained nodes and networks. CoAP is an IETF standard defined in RFC 7252 and, while having roots in the well-known HTTP protocol used in web-based applications, it has been adapted for the unique needs and restrictions of small, low-cost IoT devices. CoAP is protocol well-suited for machine-to-machine applications such as building automation and controlling smart lighting or energy. CoAP is typically run over UDP with DTLS (Datagram Transport Layer Security), however, it has been modified to run over TCP or even a WebSocket as defined in RFC 8323.

Like MQTT, CoAP has the concept of clients (or nodes) and server(s). Nodes do not directly connect to each other, rather they may connect to a single server or even maintain multiple simultaneous connections to different servers. In the CoAP model, the node becomes the center of the topology, as opposed to the server as previously described with MQTT.

CoAP uses a RESTful API to interact between a node and server. An Application Programming Interface (API) defines how software components should interact. It clearly defines a framework for the type of calls or requests, how to make them, and the data formats that should be used. An API declares an interface to interact with its logic without having to know about everything that is going on behind the curtain. Keep in mind an API is more than just an interface, it is also the specification and format. Fortunately, APIs can be designed based on an industry-standard to ensure interoperability such as OpenConfig. APIs can also be designed to be entirely custom for each vendor platform or for specific components.

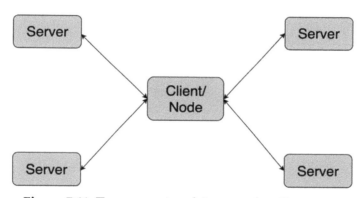

Figure 7.11: Transparent and Aggregating Gateways

Endpoints aggregate features and functions together with the purpose of exposing these functionalities. Each endpoint has a specific format for both requesting and receiving information. Specifics of any given API format can usually be found in that API's documentation. If you are familiar with the concept of functions, think of endpoints as a single function or even a function composed of many functions; you don't need to know what is going on in the function. If you properly format your request, you will be able to consume (or use) the endpoint in your application.

Postman is a popular application used to test HTTP calls to the API. Perhaps more importantly, you can observe the format of the data consumed, often returned in a JSON or XML format.

CoAP does offer secure connectivity over an encrypted path. Without encryption, CoAP uses UDP port 5683 by default. DTLS is used for the secure connection using port 5684 by default, sometimes called CoAPS.

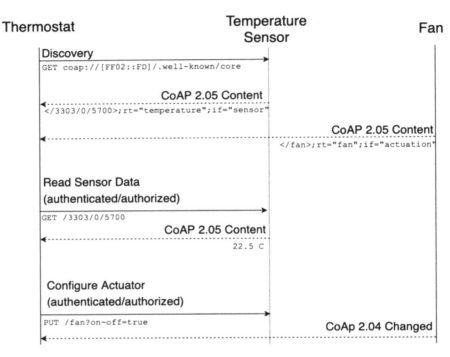

Figure 7.12: Example CoAP Communication Flow

Both MQTT and CoAP are well suited for low bandwidth and low power devices. Both protocols are frequently used in IoT applications and are capable of transmission across a secure link.

MQTT does not have a concept to alter information on the opposite node, it merely transfers information between nodes. A message emulating the function of a GET, PUT, POST, or DELETE would need to be manually defined on the MQTT publisher. MQTT natively supports QoS and retain capabilities. This is superior to CoAP, as it allows MQTT to operate when both nodes are not connected at the same time or across a link with intermittent connectivity. CoAP could mirror this functionality, but it would require you to manually program as it is not part of the protocol definition.

Hypertext Transfer Protocol (HTTP) and HTTPS (Secure)

HTTP is a well-known protocol used on the Internet for access to websites, downloading files, and other tasks. It can also be used as the IoT Data protocol within an IoT solution.

HTTP defines methods (sometimes referred to as verbs) to indicate a desired action to be performed on a specified resource. The action could be a request to list the status of all connected lights, or it could be an instruction to close a door. There is no real limit to the number or type of actions that could be created. You simply pass the parameters that define the action in a standard HTTP method, such as POST, GET, or PUT.

The five most commonly used methods with HTTP are described below:

- POST – always use POST for CREATE operations (non-idempotent)
    - HTTP POST http://www.IoTportal.io/users
    - HTTP POST http://www.IoTportal.io/users/cwnp/nickame
- GET – use GET for READ operations (idempotent)
    - HTTP GET http://www.IoTportal.io/users/cwnp/nickname
    - HTTP GET http://www.IoTportal.io/devices?size=100&page=3
- PUT – always use PUT or UPDATE operations (idempotent)
    - HTTP PUT http://www.IoTportal.io/users/cwise
    - HTTP PUT http://www.IoTportal.io/users/cwise/nickname/IoTguru
- DELETE - use DELETE for DELETE operations
    - HTTP DELETE http://www.IoTportal.io/users/cwnp
    - HTTP DELETE http://www.IoTportal.io/users/cwise/nickname/IoTguru
- PATCH – use to make a partial UPDATE operation (non-idempotent)

You may be asking what idempotent and non-idempotent means. The true definition involves a long philosophical discussion, but to oversimplify it, think of idempotent as interacting with the data set and not modifying any resources by doing so. By reciprocation, non-idempotent implies interacting with the data set does modify the resource. If you are familiar with Schrödinger's Cat thought experiment – a cat inside the box is both alive and dead. Observing the state of the cat would be non-idempotent because once observed the state of the cat is modified.

When a request is made of an HTTP server (directly through a web browser or through an application using REST APIs), the server will respond with HTTP respond codes.

These response codes indicate success, failure, security problems, etc. HTTP supports five classes of response codes:

- Informational (100-199)
- Successful (200-200)
- Redirects (300-399)
- Client Errors (400-499)
- Server Errors (500-599)

The following codes Informational response codes should be understood:

- 100 – Continue – Everything is OK so far.
- 101 – Switching Protocol – Sent as a response to an upgrade request, like switching to WebSockets.

The following codes Successful response codes should be understood:

- 200 – OK – The request succeeded.
- 201 – Created – The request succeeded, and a resource was created, usually a response to a POST or PUT.

The following codes Redirects response codes should be understood:

- 300 – Multiple Choice – The request as more than one possible response and the user or agent must choose the desired option, though the protocol does not specify how this is done.
- 301 – Moved Permanently – The URL requested has been permanently changed and the new URL is provided in the response.
- 302 – Found – The URL requested has been temporarily changes and the temporary URL is provided in the response.

The following codes Client Errors response codes should be understood:

- 400 – Bad Request – The request was not formatted in such a way that the server could understand it. Typically, the result of invalid syntax in the request.
- 401 – Unauthorized – The server requires authentication, and the client is not authenticated.
- 403 – Forbidden – The client lacks permissions to access the content even though the client identity is known.

- 404 – Not Found – The requested resource is not found on or by the server.
- 405 – Method Not Allowed – The request method is not allowed because it has been disabled. The DELETE method is sometimes disabled. Only the GET and HEAD methods cannot be disabled.

The following codes Server Errors response codes should be understood:

- 500 – Internal Server Error – The server encountered some internal problem that it cannot solve.
- 501 – Not Implemented – The request method is not supported by the server.
- 503 – Service Unavailable – The server cannot handle the request at this time. It may be down for maintenance, or it may be overloaded.
- 505 – HTTP Version Not Supported – The HTTP version used within the request is not supported by the target server.
- 511 – Network Authentication Required – The client must authenticate to gain network access.

It is not unusual, when working with REST APIs to call on an API and receive a success code (200), but no data is returned. This can occur because the HTTP server is working and it has responded to your request properly, but the server code that instantiates the REST API is not working properly and is unable to retrieve the desired information. The API may not be passing a message through the HTTP server to indicate this reality. So, you see a 200 message, but the API isn't working correctly.

The other issue often faced with REST APIs is that you receive a 403 response code when calling the API. It is possible that an API solution can be built that allows an individual to access some APIs but not others. The 403 response code indicates that the authentication token is incorrect for the API call. In other words, you don't have access.

These response codes become extremely important when working with REST APIs that are built upon HTTP communications.

REpresentational **S**tate **T**ransfer (REST) refers to a software architectural style with a set of constraints. REST is defined by six architectural constraints to make any web service a truly RESTful API. Those six constraints are:

- **Uniform interface** – resources across the system should follow the same naming conventions, link formats, or data formats (HTML, XML, or JSON)

- **Client-server** – client and server MUST be independent
- **Stateless** – each request is independent without storing session or history
- **Cacheable** – improve scalability and performance by caching when possible
- **Layered system** – segment components of the system into smaller manageable chunks
- **Code on demand** (optional) – rather than returning data, integrators may return executable code

Typical response messages from HTTP requests come back in Hypertext Markup Language (HTML), Extensive Markup Language (XML), or JavaScript Object Notation (JSON). Defined constraints along with defined message responses establish a web service framework. A web service that conforms to the REST architectural style is said to be a RESTful web service. RESTful systems provide fast performance and reliability.

RESTful API concept is widely used for web services. Many systems offer web portals with an API interface. RESTful calls are easily recognizable looking at the URL in your web browser. It usually starts with the web address of the service, some useful information that defines the service you want to use and then a question mark with some more references. The example below searches twitter for all messages posted by @CWNP.

```
https://api.twitter.com/1.1/search/tweets.json?1=%40cwnp
```

Of course, for this to work, you must be authenticated. In the case of Twitter, it means you must apply for and receive a developer account. The following image shows the result of calling this (which can be done in a Web browser) without proper authentication. The image shows a 400 HTTP response, which means that it is a bad request. Of course, this isn't very helpful. Thankfully the Twitter API is well coded, so we get meaningful data with response code 215 in JSON (this is the Twitter API and not the HTTP response code), which means Bad Authentication.

7.2: Data Storage and Streaming Plans

The preceding section explained some of the most common IoT Data protocols. Those protocols are used to get the data to the desired location. However, what happens to the data after it gets there? This is answer in the data storage and streaming plans. Once received, the data may be immediately processed, and this would be a streaming solution. If the data is immediately stored after reception, this would of a storage plan. Of course, if the data is both processed and concurrently stored, you could call it a storage and streaming or streaming and storage plan. First, we will cover storage options and then we will briefly address streaming.

Data Storage Options

A database management system (DBMS) is an application or service that runs on a computer (or cluster of computers) and provides access to data. Several types of DBMS solutions exist. We have structured databases (usually called a relational database), NoSQL databases, Document Store databases, and even big data databases.

The *database model* defines the way in which the data is stored. Many modern databases use the relational model, but other models also exist. In general terms, the database model is the type of database. Two primary types are still in use today that have been used for decades: flat-file and relational databases.

All of the information in a *flat-file database* is stored in a single storage container. Table 4.1 illustrates a flat-file data structure.

IDevID	IDevDesc	IDevLoc	Sensor1	Sensor2	Sensor3
1021	Monnit	Plant Floor	45	176.3	49
1032	Monnit	Front Lot	37	134.8	51
1021	Monnit	Plant Floor	45	176.5	50
1032	Monnit	Front Lot	37	134.7	50
1021	Monnit	Plant Floor	44	135.2	49
1032	Monnit	Front Lot	38	135.1	49

Table 7.1: Flat-File Data Structure

Here are a few key points to consider regarding flat-file databases:

- **Flat-file databases may result in high levels of data redundancy.**
  If you examine Table 7.1, you can see redundancy in action. Note that the IDevDesc (device description) is repeated for each line item, as well as the IDevLoc (device location). If a separate table were used to store the device information, this redundancy could be avoided. For example, another table could be created with IDevID, IDevDesc, IDevLoc (and any other values that are static and describe the device) rather than repeating these values every single time an entry is made to the database.

- **Flat-file databases may cost more when data is added.**
  Because flat-file databases result in more redundancy, the system simply must write more information when data is added. It is rewriting the same information again and again. When referring to an information system, the term *cost* can mean dollars and cents, or it can mean resource costs (CPU, memory, and so on). In this case, the costs are resource costs. You cannot ask a system to do more without consuming more resources within that system. In some cases, this cost is more than made up elsewhere and a flat-file should be used, but this must be considered.

- **Working with flat-file databases may be easier for some users.**
  This point is actually a positive characteristic of flat-file databases, and it is one of the many reasons you create views in relational databases. Flat-file databases are often easier for users to work with because all the data is in one location. Consider the two SQL statements that follow. Don't worry if you don't fully understand SQL yet, you will learn more about it later in the chapter. Although the increased complexity of the relational database query may seem trivial, consider what it might look like if you must join five or more tables together to

retrieve the needed information. Because all of the data is in a container in the flat-file format, no join statements are needed, and all of the data is easily accessed by decision-support professionals or business managers who may not understand the complexities of relational queries.

```
--This first query is on a relational database
SELECT dbo.Products.ProductID, dbo.Products.ProductName,
       dbo.Sales.OrderID, dbo.Sales.Quantity, dbo.Sales.Price
FROM dbo.Products
INNER JOIN dbo.Sales ON dbo.Products.ProductID =
dbo.Sales.ProductID;


--This second query retrieves the same information from a
flat-file database
SELECT dbo.Sales.ProductID, dbo.Sales.ProductName,
       Dbo.Sales.OrderID, dbo.Sales.Quantity, dbo.Sales.Price
FROM dbo.Products;
```

This simplification is one of the driving factors behind many views that are created and behind many of the decisions that are made when online analytical processing (OLAP) databases are implemented. OLAP databases are usually read from (far more read operations are performed as opposed to write operations), and they may benefit from a flattened model; however, even with OLAP databases, it is still common to have multiple tables. The tables may simply be less *normalized* (understood as more redundant) than those for an online transaction processing (OLTP) database that processes large numbers of writes to the data.

The concept of a flat-file database is expanded with NoSQL databases to bring it into the modern world. NoSQL databases do not equal flat-file databases, but they could be said to be an evolution of them in a different direction from relational databases.

Structured (Relational) Databases
Relational databases store information in separate containers called *tables*. Each table represents a single entity, although *denormalized relational databases* may not always do so.

Normalization is the process used to ensure that a database is properly designed according to relational database principles. The term is best understood by considering what you are attempting to avoid as a database designer or administrator. Common database problems can be all but eradicated through the process of normalization. Problems that present themselves in poorly designed relational databases include duplicating data, creating improper data relationships, and constraining data entry. They must be addressed either during the design process or through repairs after the database is in production. Needless to say, it is much easier to deal with the problems before the database is actually implemented. You can think of these problems as abnormalities or as anomalies. Just a few of these problems are outlined in the following list:

- **Duplicate Data:** Duplicate or redundant data results in poor database performance for many databases. For example, if a database is used to store sales information and the entire customer information set is stored in each record for each sale, the result is a much larger database. Not only is the database larger, but each record is larger than necessary, too. This increased record size causes slower writes in every situation, and it may cause slower reads as well.
- **Improper Data Relationships:** When unrelated data is stored in a record, managing the data becomes difficult. If you want to delete the customer record but not the sales information referenced in the preceding paragraph, you would need to manually process each and every record to delete all instances of a given customer. This work is not only time-consuming but introduces an increased likelihood for error. The problem is born of improper data relationships. The customer is not the sale, and the sale is not the customer. They should be stored separately in an online transaction processing (OLTP) database because such databases are usually more normalized so that writes and changes are optimized.
- **Data Entry May Be Constrained:** One common mistake in relational database design is the implementation of table structures that include columns such as Item1, Item2, Item3, and so on. What happens when the data entry employee needs to enter a fourth item into a table that stops at Item3? The situation cannot be accommodated unless the data entry employee enters multiple items in a single column, which results in the loss of data integrity.

These problems, and others not listed, can be categorized as three potential anomaly types:

- **Insert Anomalies:** An insert anomaly usually occurs when improper data relationships are designed within a single table. An example would be a sales table that contains the customer ID, customer name, and customer address information, while no separate customers table exists. You cannot add a new customer to this database without entering a sale into the sales table. This is an insert anomaly.
- **Delete Anomalies:** Delete anomalies are the mirror of insert anomalies. Continuing with the preceding example, if the last sale for a given customer is deleted, all information about that customer is also deleted. You could delete the information about the sale only if you specifically deleted the sales column values while leaving the customer column values. Because the table is a sales table, the result would be an orphaned sales ID value used only to locate the customer information.
- **Update Anomalies:** Update anomalies are another result of redundant information entry. If a customer calls and informs the data entry employee that a new address should be associated with the company, every sales record must be updated. If only one sales record is updated, future sales could be shipped to the wrong address.

These problems and anomalies are possible because the data is not stored in the most effective way. To help database designers implement better tables, Dr. E. F. Codd, a systems engineer and scientist heavily involved in database system solutions during the 1960s and 1970s, defined three normal forms: first normal form, second normal form, and third normal form. Over the years, several other normal forms have been developed as well. However, most database designers target the third normal form and then implement other, more advanced, normal forms only as they are needed.

Relational Database Benefits

When working with structured data, relational databases provide the following benefits over flat-file databases and, in most cases, NoSQL or Document Store databases:

- **Relational databases can be indexed and optimized more efficiently with structured data.**
    Relational databases can be indexed and optimized more efficiently because you are dealing with smaller units of information in each data store (each table). For example, you can index the Sensor Tracking table uniquely for retrieving common columns of information, and you can index the Device Description table

uniquely for retrieving common columns of information retrieved from that table. If the two tables were crammed together into a single flat structure, you would have to ask which is more important: Sensor Data columns or Device Description columns (of course, to me the answer is obviously Sensor Data columns most of the time, but you understand the point). You can create only so many indexes before you start hurting more than you help.

- **Relational databases consume less space to store the same information than flat-file databases when storing structured data.**

    Because the redundancies have been removed, a relational database requires less space to store the same information as a flat-file database. For example, consider Table 7.1 again. The IDevID, IDevDesc, and IDevLoc fields must be added every time a specific device produced data for storage in the database; however, with the structure in Figure 7.13, only the IDevID must be added with each record in the sensor tracking table (upper right) and the IDevDesc and IDevLoc values are stored only once for each device. You are, therefore, dealing with one column instead of three. You can see how the relational structure saves on storage space. Now imagine you have 1000 sensors reporting. You can see the significant different this would make.

- **Relational databases can handle more concurrent users more easily with structured data.**

    Because data is broken into logical chunks, relational databases can handle more concurrent users more easily. With the data store represented in Table 7.1, even if the user wants only the sensor reading-specific information with no information about the device, all the data must be locked in some way while the user retrieves the information. This behavior prevents other users from accessing the data, and everyone else must wait in line (what a database system usually calls a *queue*). The relational model is better because one user can be in the Device Description table while another is in the Sensor Tracking table. Of course, modern database systems go even further and usually allow locking at the data page or even the row (record) level.

- **Relational databases are more scalable when working with structured data.**

    Because they allow for more granular tweaking and tuning, relational databases *scale* better with structured data. They store more information in less space. They allow more users to access the data more quickly.

I hope you caught the pattern throughout this section. I continually clarified that the data was structured. This is the key characteristic that makes relational databases perform well. If it is unstructured, you should evaluate the performance results you will get with unstructured database systems.

Of course, the fact remains that a relational database that is heavily normalized (with extreme reductions in redundancy) may be much more difficult for users to utilize. For example, it is not uncommon to see the typical report records built from four or more underlying tables in modern relational databases. This structure means that the users must join the four or more tables together to retrieve that typical report records desired. One of the key decisions a DBA makes is determining just how normalized a database needs to be.

IDevID	Sensor1	Sensor2	Sensor3	SRecID
1021	45	176.3	49	1000
1032	37	134.8	51	1001
1021	45	176.5	50	1002
1032	37	134.7	50	1003
1021	44	135.2	49	1004
1032	38	135.1	49	1005

IDevID	IDevDesc	IDevLoc
1021	Monnit	Plant Floor
1032	Monnit	Front Lot

IDevID	IDevDesc	IDevLoc	Sensor1	Sensor2	Sensor3
1021	Monnit	Plant Floor	45	176.3	49
1032	Monnit	Front Lot	37	134.8	51
1021	Monnit	Plant Floor	45	176.5	50
1032	Monnit	Front Lot	37	134.7	50
1021	Monnit	Plant Floor	44	135.2	49
1032	Monnit	Front Lot	38	135.1	49

Figure 7.13: Sensor Tracking with Relationships

Important Relational Database Terms

As you learn about programming and relational databases, you will encounter many terms related to database implementation and management. It is important that you understand the definitions for these terms as used in this book and others. Many terms have more than one definition, and it is important that you understand the meaning poured into the words in context. Some of these terms are basic, and some are more complex, but you will see them appearing again and again throughout this book and as you read articles, white papers, and websites related to the work of a database administrator or design professional. The following list will define these common terms used in the knowledge domain of relational databases:

- **Table/Record Set/Relation**
  In relational database design, a table is not something at which you sit down to eat. Rather, a *table* is a container for data describing a particular entity. Tables are sometimes called record sets, but the term *record set* usually references a result set acquired by a SELECT statement that may include all or a portion of the table data. The formal name for a table is a *relation*. All the entries in the table are related in that they describe the same kind of thing. For example, a table used to track motor actuators describes motor actuators. All entries are related to motor actuators.

- **Column/Field/Domain**
  To describe the entity represented in a table, you must store information about that entity's properties or attributes. This information is stored in *columns* or *fields* depending on the database system you're using. For example, the Motor Actuators table would include columns such as DeviceID, Model, SerialNum, Location, Description, and Function, possibly. Note that these properties all describe the motor actuator. The term *domain* is used to reference a type of property or attribute that may be used throughout the database. For example, you may consider DeviceID, Model, and SerialNum to be domains. To ensure domain integrity, you would enforce the same data type, constraints, and data entry rules throughout the database for these domains.

- **Record/Row/Tuple**
  A collection of columns describing or documenting a specific instance of an entity is called a *record*. Stated simply, one entry for a specific unit in the Motor Actuators table is a record. Records are also called rows in many database systems and by many DBAs. The formal term for a record is a *tuple* (usually pronounced "too-pel," but some argue for "tyoo-pel").

- **Primary Key/Foreign Key**
  The primary key and foreign key play an important role. The primary key is the unique identifier for each record in the table. For example, in the Motor Actuators table, DeviceID could be the primary key. Only one record in that table, no matter how many millions or billions of records might exist will have a particular DeviceID value. Now, let's say you have another table where you track departments to link devices to departments. You might have a primary key

in the Departments table called DepartmentID. In the Motor Actuators table, you could reference all of the details about the department to which a particular actuator belongs by simple adding a DepartmentID column to the Motor Actuators table and entering the value of the proper department. This new value is called a foreign key. The foreign key relates the two tables together or forms a relationship.

- **Index**

  An *index* is a collection of data and reference information used to locate records more quickly in a table. For now, it's enough to know that an index can be used to increase database performance and that they can equally decrease database performance when used improperly.

- **View**

  One of the most over-explained objects in databases is the view. Here's the simple definition: a *view* is a stored SQL SELECT statement. That's really all it is. Views are used to make data access simple, to abstract security management, and to improve the performance of some operations. The most common use of views is the simplification of data access.

- **SQL**

  SQL is the database communications language managed by the ANSI organization. It is a vendor-neutral standard language that is supported at some level by nearly every database product on the planet. Most DBMS solutions will implement standards-based SQL today, for the most part, and then add extra features to it as required to form their own "flavor" of SQL.

- **Stored Procedure/Embedded Code**

  When you want to process logical operations at the server instead of the client, stored procedures can be used. A *stored procedure* is either a collection of SQL statements or a compiled stored procedure, if supported by the DBMS. Basically, you have internal stored procedure and external stored procedures where the DBMS calls on an external application to perform some operation. Many DBMS solutions have enhanced versions of SQL that provide logic so that internal stored procedures are more powerful than standard SQL alone. Stored procedures are used to centralize business rules or logic, to abstract security management, or to improve performance. Other reasons exist, but these are the big three motivators.

- **Trigger**

  A trigger is like a dynamic stored procedure. A *trigger* is a group of SQL statements that is executed automatically when specified events occur. For example, you may want to launch a special procedure anytime someone attempts to execute a DROP TABLE (delete a table) statement. The trigger could either back up the table before deleting it or simply refuse to delete the table.

- **Concurrence**

  Concurrence is defined as acting together. In the database world, either a system supports multiple concurrent users, or it does not. *Concurrency* is a single word that says a database system supports multiple users reading and writing data without the loss of data integrity.

- **Data Warehouse**

  A *data warehouse* is a special kind of database used to store large collections of data (like a real warehouse). They are frequently used to store older data that is not required as frequently as more recent data. They server other purposes as well.

- **DBA**

  A *DBA* is a database administrator. A DBA is the person who installs the IoT devices and gateways, implements the authentication solutions, builds the user databases, configures the control applications, programs the PLCs, troubleshoots production and security problems, and, oh yeah, works with databases on occasion. But seriously, you live in a new world of IT/OT. Today, many IT/OT professionals must wear multiple hats. This reality means that DBAs usually must know about the database server service, the server operating system, and even a bit about the network infrastructure across which users communicate with the database system. It's a brave new world. In IoT, it's becoming even braver.

NoSQL and Document Store Databases

Before explicitly defining a NoSQL or Document Store database, let's revisit an earlier topic. I said that flat-file databases may be slower with write operations. This is true when the data could be stored in a relational format. But what if the data is not structured to allow for this? What if the data varies significantly from write-to-write? Sometimes it includes eleven fields of information and sometimes it includes only three. Now, we need a new solution. We need an evolution of the flat-file concept that allows for the storage and access to and of unstructured data with fast writes and fast reads. This is where NoSQL and Document Store databases come in.

Fundamentally, a NoSQL database is simply one that is not modeled after the relational table structure used in relational databases[213]. According to the mongoDB website, one of the leading NoSQL databases in the Document Database categories, "NoSQL databases (aka "not only SQL") are non-tabular, and store data differently than relational tables. NoSQL databases come in a variety of types based on their data model. The main types are document, key-value, wide-column, and graph. They provide flexible schemas and scale easily with large amounts of data and high user loads.[214]"

While NoSQL databases do not always use tables, they do support relationships. The relationships are object-based relationships. Consider a document database where you have a collection (a group of documents of the same type) for devices and a collection for reported sensor data. You can relate a device to sensor data by storing the devices collection document ID in the sensor data documents linked to that device. When you do this, the device's data (like description or location) from the document in the devices collection is stored in line with each sensor data document. Then, if you change the location of a sensor device in the devices collection document for it, the Document Database is intelligent enough to also update the document linked (related) to that device.

These documents we're referring to are most often JSON documents. So, they are documents in the sense that they can contain whatever information you want and each document can contain some information that may not be in others, but we typically design these databases with a logical structure behind them. That is some documents may have 12 items in them and others have only 6, but they all have the possibility of the same 12 items. We impose this constraint programmatically. The following is what a JSON document might look like:

```
{
    "deviceid": "sens1076",
```

---

[213] Some database systems ride on the border between relational and NoSQL. They have some features from both and don't really fit perfectly into either category. In such cases, they may be able to support structured and unstructured data well.

[214] https://www.mongodb.com/nosql-explained

```
            "description": "Laid technology temperature sensor",
            "location": "plant floor, section A",
            "department": "Receivables"
}
```

Yes, that text is what you call a document in a Document Database, so don't be confused by the terminology.

NoSQL databases come in four basic types today:

- **Key-Value Databases:** This is the simplest form of NoSQL database and it stores information in key:value pairs. Some think of it like a table with only two columns, the key, which is the name of the attribute, and a value for the key. An example of this database type is DynamoDB in AWS.
- **Document Databases:** Document databases store JSON documents in most cases, but many support other formats as well such as XML, and BSON (Binary JSON). JSON is typically stored in UTF-8 string format and supports strings, Boolean values, numbers and arrays and you can look at it and see what it means. BSON is stored in binary format and supports the same data types plus date, and raw binary, but you can't look at it and read it. However, it is much faster for computers to process. The documents in a document store database are stored in collections and a single database can have multiple collections.
- **Column-Oriented Databases:** Instead of storing data in rows, these databases store data in columns. This structure allows for very fast calculations against columnar information. Consider, for example, that you want the average temperature reading from 700,000 entries in a database. A relational database would need to retrieve all those values from all records, which would result in a lot of paging (moving around in the underlying database structure) to retrieve the information and perform the calculation. With a column-oriented database, these calculations are faster as the data is stored (as near as possible) sequentially down the column. Hence, these databases are used more in the analysis role.
- **Graph Databases:** Graph databases store relationship among the elements in the database. This is different from relational databases. Relationships in relational databases are often considered "assumed" because the actual relationship isn't always stored. But they are used to enforce integrity through processing. Graph databases are about analyzing these relationships among entities and so the

relationship exists in the database because each entity stored is a node that is "related" or "connected" to another node.

At this point, you're probably beginning to wonder, "Which one of these databases should I be using?" The answer may be, "All of them!" Or it may be the one or ones that meet your needs. The beauty of data is that it has no feelings. It doesn't care if you move it around, transform it, or even delete it. So, you can do with it what you need.

For example, you may use a document database for your first level storage, but then you might process that data with ETL and restructure it for a relational database or a graph database and use it there as well. The point of IoT data storage planning is to get the data to the needed users and systems in the way they need it. All these various database types help you to accomplish this goal.

While relational databases nearly all use SQL, in some flavor or another, which makes knowledge transfer from one relational database to another easy, this is where NoSQL databases lose the advantage. It's not uncommon to see very different query languages (often called APIs with NoSQL databases) from one database system to another. So, you will have to learn the specific query methods (and write methods to be precise) in order to work with the data.

Additionally, it is often said that NoSQL databases lack support for complex queries. To be clear, this is generally true. However, because the queries are executed with APIs, it's up to you how you process the data returned. The problem with lack of complex queries inherent in the system is that you may have to return more data (so that you can process it in code yourself) than you would have to with a complex SQL statement. In the end, it is true that NoSQL databases generally support less complex queries than relational databases.

When it comes to IoT data, it varies significantly from one IoT solution to another. Some IoT solutions (sensors and actuators and other devices) do generate structured data. In such cases, relational databases may be better; however, if you have thousands of these devices and they send in data hundreds of times per day, even the structured data might be better stored in a NoSQL database as you will see in the next topic on big data. Other IoT solutions actually generate unstructured data or, at least, inconsistently structured data, which may also be best stored in a NoSQL database.

In the end, if the data generated by the IoT devices is unstructured, it's probably best to go with a NoSQL database – at least for the initial data storage. Also, if you are generating more than 15-20 gigabytes of data per day, you should probably take a serious look at NoSQL solutions.

One common use of NoSQL is big data. While the storage mechanisms used in NoSQL will require more storage space for the data, with the right optimizations it can handle larger amounts of data better than relational databases (think, exabytes of information or more).

Gartner defines big data as huge-volume[215], fast-velocity, and different variety information assets that demand innovative platforms for enhanced insights and decision making. In other words, it's a lot of data coming in fast with significant variations in the structure of the data.

Generally speaking, today, we consider a data set or system big data if it must deal with terabytes of data or more as opposed to gigabytes of data or less. As time goes on, this will change, but it is the current thinking of most big data experts.

You could also define big data in a more general sense as:

> *Big data consists of data sets that are too large or complicated for traditional single-server relational database systems to process well.*

Because big data is so, well, big, it is typically processed using clusters of servers instead of single servers. A single server has difficulty meeting even one demand against such a large data set much less multiple concurrent demands. This is true whether the demand is writing new information as it streams in from thousands of devices or if it is reading information from the database by multiple users.

One of the most popular big data platforms is Apache Hadoop. Hadoop implements the High Distributed File System (HDFS) as the storage layer. All files stored in HADOOP are implement as block-sized chunks and stored. Within HDFS, two nodes (servers) exist:

---

[215] Did you know that Facebook generates more than 500 terabytes of data per day and that Twitter, even with its constraints on input size, generates more than 8 terabytes of data per day? That's big data.

NameNode (the metadata about block locations is stored here) and the DataNodes (the actual block storage location, which is controlled by the NameNode).

Hadoop uses MapReduce as the processing layer. MapReduce provides the framework used to write applications with distributed processing, like Hadoop.

Hadoop can be scaled horizontally (adding nodes) and/or vertically (using more powerful nodes or nodes with more storage). Additionally, resiliency is provided through HDFS as it uses replication to store blocks on multiple nodes (by default 3 nodes) or (in newer versions) it uses algorithmic processes to get the same level of resiliency with less storage overhead.

Ultimately Hadoop is a highly distributed database system using multiple nodes operating in parallel on the data stored within HDFS, while each node operates on its own local data.

Data Streaming

The typical batch storage processing pipeline is represented in Figure 7.14. It begins with data acquisition and ends with the data stored and ready for serving. Along the way, the data has been transported, stored, and processed.

The acquired data will be in some format, communicated on some interface, possibly secured through authentication to the interface or with encryption of the data, generated with some level of reliability (based on the sensor accuracy, for example) and with some level of latency (the rate of telemetry data generation at the sensor device).

It is then transported to the storage server (DBMS server) with some level of reliability in transport (not all systems require that one hundred percent of the telemetry day reach the server), integrity of the data transported, at some level of latency and cost of delivery.

Next, the data is stored within the DBMS. Keep in mind that, between the transport and storage, an IoT Data protocol may exist. That is the acquired data is transported to an MQTT server, for example, and then a subscriber application reads the data from the MQTT server and has a connection to the DBMS where it can write the data. The storage system will offer some level of flexibility and schema requirements (SQL will be less flexible and have a constraining schema and NoSQL will be more flexible and have no schema or a less-constraining schema), some query options for access to the data (SQL or

MQL (mobgoDB), for example), with levels of availability, reliability, and cost, depending on the design of the DBMS implementation.

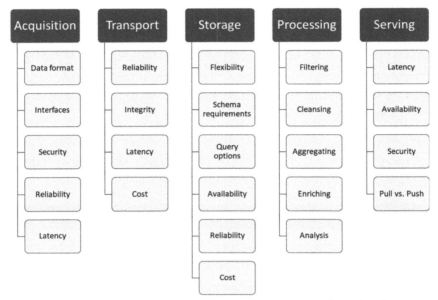

Figure 7.14: Common Batch/Storage Data Processing Pipeline

At this point, we can say that the data is in initial storage. It may or may not be in the format required by all applications. If it is not, an intermediate layer between storage and serving would exist for processing. Here the data can be filtered (limited to a specific subset of the data), cleansed (increase data accuracy, increase data completeness, remove data duplicates, convert data to appropriate formats (for example, from one temperature format to another)), aggregated (grouping data from multiple stored data sets), enriched (add data from outside sources, such as Internet weather sites, health sites, etc.), and analyzed (new data generated based on the resulting data from the previous stages). The processed data may be stored in different DBMS systems or back into the same system in different databases or tables.

Finally, we're ready to serve the data to the business applications. The applications can access the data with some level of latency, availability, and security. The data may be accessed by request (pull) or by active delivery (push). You can see that this traditional pipeline includes a lot of variables and much to be planned. However, in some cases, we cannot wait for all this initial storage, processing, and serving. We need to cut out those last three functions. This brings us to data streaming.

It is common in IoT solutions to discover the need to process streaming data. If an IoT device sends a continual stream of information, it falls into this category. Don't confuse this with the general category of streaming protocols, like MQTT, as those protocols are not always transmitting one frame after another frame after another (although they can). In this case, rather than referencing the general data communication method, we are indicating that the device is continually sending data with no significant intervals between transmission.

Examples of streaming include video and audio as very common examples. However, it also includes mission critical control systems that must report status multiple times per second to gain response back from the control system before a severe problem occurs. This is not uncommon in plant automation systems. It is one of the reasons that HART is popular as a protocol because it has an always on, 100% of the time detector as to whether a HART device is there or not. That detector is the fact that power is always flowing in the wire even if the device is not transmitting. The only way to accomplish something similar with wireless IoT would be to transmit very rapidly and, to be direct, its one reason that wireless IoT is not always the best choice.

Now that you see the intention of streaming data in this context how do you deal with it? The best answer is in memory and not in storage. For example, if a plant control device reports status twice per second and needs to receive a stop signal within two seconds, if required, you do not want to have the added delay (yes, even with Hadoop) of storing the data and then having it analyzed by another process before a response is given. Instead, an in-memory process should evaluate the data before it is written to the database and respond to the device before taking the time to write to a database. After responding, it can store the data in a database, if required.

When implemented appropriately, the storage may occur concurrent to the processing. Consider the following example. The data is sent in a stream to an application server. The control system received the data as it comes into the application server. At the same time a data writer application is reading the data from the application server and transferring it to storage as required. Such a plan can accomplish concurrent or near-concurrent data analysis and data storage. This model is represented in Figure 7.15 with the alternate flow. That is, we may choose to process the streaming data through in-memory serving of that data and then place it into storage if necessary. Alternatively, we may choose to implement both immediate serving of the in-memory data to applications for real-time processing and concurrently pass the data off for storage processing (usually handled by another server or service).

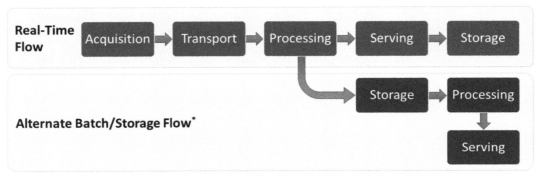

Figure 7.15: Data Processing Pipeline for Streaming and Storage

That's the first part of the answer. The second part goes further. You also have to determine if you want or need to write every transmission from the device to the database. You may choose to retain the 120 most recent transmissions in memory at all times (that would be 1 minute's worth at the rate of twice per second) and write every 60th transmission to the database. What keep them in memory if you're not writing them to the database? In the event of the need for a stop command, you may choose to write all in-memory transmission to the database leading up to the one requiring the stop command and all future transmission until an administrator disables it. This gives you information around the time of failure to analyze.

How will you make all these decisions? Remember, it should be in your requirements. That's why requirements are so important.

When it comes to storing the stream of data, your database choice will depend on the kind of data, but NoSQL is often a good choice.

7.3: Application Planning

Application planning introduces many concepts including cloud, fog and edge processing, containers and container orchestration, and general business applications. We will briefly explore all three in this section so that you can provide recommendations on their use and planning.

Cloud, Fog, and Edge Processing

We have discussed the cloud, fog and edge processing issue previously in the book and will not cover it again in-depth at this point. However, emphasis must be placed on the constraints that drive our decision among these choices. Figure 7.16 illustrates the concern.

First, we can classify processing tasks in many ways. Figure 7.16 represents a model where tasks are classified using the following categories:

- Long-Term Use
- Delay Tolerant
- Low Latency
- Extreme Low Latency

If we define these classifications based on time-to-completion or time-to-response, we can define them in seconds or milliseconds. For example, we may define them as follows:

- Long-Term Use: Acknowledge receipt of data; no time constraint.
- Delay Tolerant: Acknowledge receipt of data; provide guidance; response within 5 seconds.
- Low Latency: Acknowledge receipt of data; provide guidance; response in less than 1 second.
- Extreme Low Latency: Acknowledge receipt of data; provide guidance; response in less than 100 ms.

These are intended as examples, but you can see how the tasks can be classified and the result will be an easier decision as to where processing should occur. In our case, those tasks classified as Long-Term Use can easily be sent to and processed in the cloud. There is not need for an urgent response with these tasks and, in all likelihood, the IoT end device will not need a response to continue operations or make any decisions.

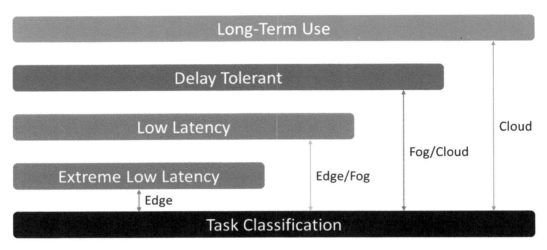

Figure 7.16: Task Classification and Processing Locations

The Delay Tolerant classification does need a response, but the response is not urgent. That is, the situation requiring guidance is not health-critical or critical to the stability of the system or physical entity. Depending on the scenario the processing may occur in the fog or the cloud.

For the Low Latency and Extreme Low Latency tasks, they are trending more toward the fog, or the edge, and the Extreme Low Latency task will likely require at least some edge processing.

The point of this model is not to adopt the classifications listed here but rather to understand the process of classifying tasks based on their requirements in relation to time and then ensuring the processing occurs such that the time constraints can be accommodated.

Containers and Container Orchestration

A common method used for application deployment today is that of containers. Containers allow you to separate your applications from your infrastructure so you can deliver software quickly[216]. They provide a standard unit of software that packages up code and all its dependencies, so the application runs quickly and reliably from one computing environment to another[217].

The modern concept of containers was born out of the heavy use of virtualization that began in the late 1990s and early 2000s. I remember running virtual machines in Microsoft Virtual PC in the early 2000s and it was a fabulous technology. However, while the technology was really cool, and still is today with VirtualBox, Hyper-V, and VMware, among others, it does not answer all problems. For example, what if I want to run an isolated application, but I do not want to run an entire virtualized operating system? The application runs on my operating system, but I want it to be easily installed, configured, and loaded and I do not want it to permanently impact my localized operating system install. This is where containers come into play.

Containers are different from virtual machines (VMs) in several ways. First, they do not require a hypervisor that fully "emulates" or "passes-through", whatever the case may be, the hardware. Instead, containers use an abstraction layer between the applications and the host operating system. So, while VMs abstract the hardware, containers abstract the host operating system. The key difference here is that containers are much smaller because you do not install the entire operating system for each container like you do for each VM. The containers share the same kernel.

Second, containers isolate processes whereas VMs isolate entire machines, though they are virtual. Each container may be isolated such that it cannot communicate with other containers or applications running on the same system. At the same time, they can be allowed to communicate with those same systems. This provides more flexibility.

These differences are illustrated in Figure 7.17.

---

[216] docs.docker.com/get-started/overview/
[217] www.docker.com/resources/what-container

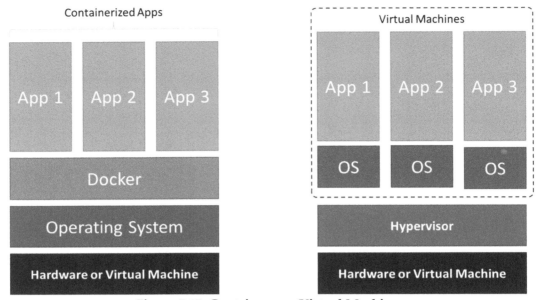

Figure 7.17: Containers vs. Virtual Machines

Containers are used frequently in IoT solutions. They may run on-remises or in the cloud. They may be instantiated only when required or run one hundred percent of the time. One of the key benefits not referenced above is scalability. If I have multiple servers running the proper containerization services, I can instantiate additional copies of a container as needed. For example, during some periods a single container may be able to keep up with the incoming IoT data, but in other periods I may require three containers on three different servers to keep up with the workload. In such cases, the additional containers can be instantiated and then removed when no longer required.

One of the keys to this dynamic use of containers is container orchestration. Container orchestration acts as a sort of middleware layer for the management of containers. For example, I can define that I need three copies (replicas) of container one between 8 AM and 6 PM each day, but I only need two copies (replicas) of it between 6 PM and 8 AM. Additionally, I can define that I need five copies of container two between 8 AM and 6 PM, but I need seven copies between 6 PM and 8 AM. The container orchestration service, typically managing a cluster of servers or workers, can dynamically schedule these containers to operate as needed.

The most popular container service is Docker and the most popular orchestration service for Docker is Kubernetes. All the major cloud platforms support both solutions. Additional container services include:

- Podman
- LXC (though it gets closer to virtualization)
- Windows Containers

Additional container orchestration services include:

- Apache Mesos (a cluster orchestrator that supports Docker)
- Docker Swam mode
- AWS Amazon Elastic Container Service (ECS)

With this understanding of containers and container orchestration, you should be able to understand their potential use at the IoT application level.

Another solution that is not quite in the containers category and also not the same as most conceptions of virtualization, is Kata Containers. Their stated goal is *the speed of containers, the security of VMs*[218]. Kata Containers is managed by the OpenStack Foundation, who also manage OpenStack itself, an open source cloud operating system that allows you to deploy a private cloud. To oversimplify, Kata Containers are generally slower than containers but faster than virtual machines. They are larger than containers but smaller than virtual machines.

Kata containers can be managed by Docker, Kubernetes, and OpenStack.

General Business Applications

For general business applications, we're focused on the user's needs. They don't care how the data is stored or if services are running on bare metal, hypervisors, or in containers. The just want to run an application, see the information they require, and be able to take actions they require. Such applications can be implemented using configuration and customization procedures with existing applications or through the complete development of custom applications. Deciding between the two is a factor of budget, time, abilities, and availability of off-the-shelf solutions.

---

[218] Katacontainers.io

First, is an off-the-shelf solution available that meets your needs or can be customized to meet your needs. If it works out-of-the-box, the cost of acquisition and deployment is still important. If it must be customized, the time and cost of acquisition, customization, and deployment becomes important. I've seen may cases where a customized off-the-shelf solution took more time and cost more money than a custom-developed solution likely would have.

It is a factor of budget because it can cost more to develop a custom application, or it can cost more to use an off-the-shelf application. It is not as simple as some like to thing: custom is always more expensive. This statement is simply not true. In many cases an organization has a codebase that may provide for twenty percent, forty percent, or even eighty percent of the target application's needs. Such code reuse can significantly reduce the cost of development. Additionally, many off-the-shelf applications are not sold based on a cost model that competes with custom development budgets but rather based on custom development times. That is, they compete with the time to deployment rather than the cost of deployment. This is a fancy way of saying that the off-the-shelf software can often be very expensive. When cost alone is not enough to make a decision, we move on to time and may be willing to spend more and get results faster.

When cost and time do not resolve the problem, we have to consider abilities. If you have developers on staff that can build the solution, cost will often be less (though this is not always the case). If you have to higher contractors or an outside development firm, costs usually increase. Additionally, in-house abilities or the lack thereof can impact time. Now, we see that cost, time, and abilities are interrelated.

Thinking along these lines will help in the decision process.

Custom-built solutions are often aggregates of multiple existing solutions as well. For example, you may use Grafana to provide visualization of IoT data, but customize the views as well as some of the code to more exactly meet your requirements.

In some cases, the only customization required is getting the data into the right format for the acquired solution. To do this, you may use an ETL tool, python scripts, or custom C#, C++, or other codebase to accomplish the transformation.

7.4: Monitoring and Maintenance Planning

Planning for the monitoring and maintenance of the IoT solution is essential. By monitoring, we mean monitoring the IoT solution itself. This section will provide an overview of monitoring solutions and maintenance processes.

Monitoring Solutions

Both commercial and open-source monitoring solutions are available for IoT. Custom-built monitoring solutions can be forged by combining existing technologies like SNMP, network flow monitoring, database monitoring, application monitoring, and visualization tools like Grafana.

When building custom solutions, a tool that is growing in popularity is Node-RED. Node-RED is described as a programming tool for wiring together hardware devices, APIs, and online services in new and interesting ways. It provides a browser-based editor that makes it easy to wire together flows using the wide range of nodes in the palette that can be deployed to its runtime in a single-click[219]. It is built on Node.js and supports embedding JavaScript code into the flows for customization. It also has a built-in dashboard function and is customizable through custom nodes (of which, the flows.nodered.org library contained more than 3400 at the time of writing). Figure 7.18 shows an example dashboard created in Node-RED.

Given that Node-RED can act as a subscriber to MQTT server topics, you can use it to harvest all the same data your IoT applications see (if they are using MQTT) and then report on this data. You can also implement CoAP, AMQP, HTTPS/REST, and other protocols in Node-RED for the building of a nearly complete IoT monitoring solution from a data flow perspective. It also supports listening to SNMP traps, communicating with mongoDB, SQL Server, Oracle, MySQL, PostgreSQL, MariaDB, and many other databases, and even interaction with cloud services like Azure IoT Hub and AWS IoT. Clearly, Node-RED can be used to build a custom monitoring solution even if you are not an advanced programmer[220].

---

[219] nodered.org
[220] It is a recent phenomenon to see Node-RED used in production systems for workflows. Early-on it was mostly a prototyping tool. But with continual improvements, it has shown to be a potential solution for production-ready implementations.

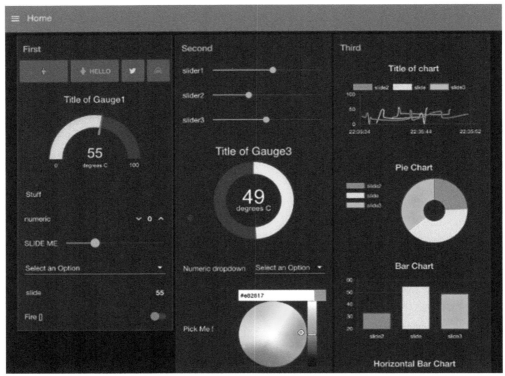

Figure 7.18: Node-RED Dashboard

Custom solutions can also be built using Python and other scripting languages. All the technologies previously listed for Node-RED can be accessed with Python. If you know the Python language, it can be used to build data harvesting and aggregation scripts, store the data in a selected database, and even build visualizations with tools like plotly and Dash.

An open-source solution, ThingsBoard, is an IoT platform for data collection, processing, visualization, and device management. Included in the toolset is IoT device and system monitoring. Be sure to consider this solution as it has proven to provide the needed features for both system monitoring and user applications in many deployments. Figure 7.19 shows the interface for adding IoT devices in ThingsBoard.

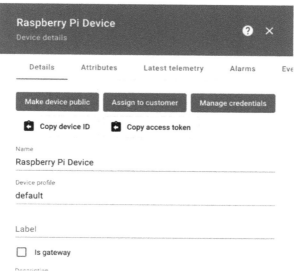

Figure 7.19: Device Management in ThingsBoard

Many commercial solutions are also available and should be evaluated to determine their cost/value relationship with a given project.

Maintenance Processes

Maintenance process planning involves the discussion of maintenance types and maintenance tasks. Maintenance types include corrective, preventive, adaptive, and predictive. Maintenance tasks include updates, backups, configuration management, monitoring, and other modifications. The maintainability of a system is the capability of a system to be modified for continued operations. Maintenance is performed on maintainable systems. Modifications can include:

- Patches
- Firmware upgrades
- Software upgrades
- Improvements
    - Added hardware
    - Added services
    - Added applications
- Adaptations (added integrations)

The maintenance types are outlined in Figure 7.20. The first two types are corrective and preventive. Corrective maintenance is dynamic. It is performed to resolve a problem that has occurred. It can be either immediate or deferred. When it is immediate, the system is modified in some way to solve the current problem. When it is deferred, it becomes adaptive maintenance. That is, a problem has been identified, but it does not prevent the system from performing its primary intended role. Therefore, an adaptive maintenance process is launched that will add integrations or other features that solve the problem or lack of an ability. Often, deferred problems are missing capabilities.

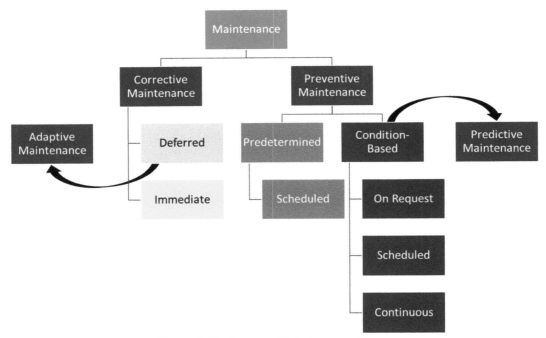

Figure 7.20: System Maintenance Types

Preventive maintenance is performed to avoid future problems. It can be predetermined or condition-based. Predetermined maintenance is scheduled and includes things like backups, hardware replacement, and database tuning. Condition-based maintenance occurs because it is triggered by an event, such as a software update release, a security patch released, or some other event that does not immediately prevent the system from function. Such maintenance may be performed on request, scheduled as a later task (converting to predetermined), or simply performed as a continual process (for example, monitoring for new security vulnerabilities daily and patching them as they occur). A condition or event can also result in predictive maintenance because it does not require immediate action but, based on the condition, you can determine when action will be required. An example would be when a system reaches 70 percent capacity, it may be a trigger alerting you to the need for more capacity within six months.

7.5: Security above Layers 1-3

In previous chapters, we've referenced security and its importance. It is equally important for IoT Data protocols, Data Storage and Streaming, Applications, and Monitoring and Maintenance. Without it at this level, you do not have end-to-end security in the wireless IoT design framework.

For the IoT Data protocols you should plan for:

- Authentication to the servers.
- Encryption of transmissions.
- Accounting/logging of actions.

Many servers will have built-in support for authentication and accounting. However, encryption may not be available as a payload function. Instead, you will have to use TLS or DTLS, as appropriate, with the IoT Data protocol in use to implement encryption. If you choose to use certificates in the process, you will also achieve a level of device authentication with these protocols.

For data storage, authentication to the database system is essential. However, you must also consider the encryption of the stored data (encryption at rest). It may be desired for an extra level of security, particularly if the server is a shared resource running other applications and services. In such case, a vulnerability in on of the other applications or services that provides an attacker with system-level access may expose your IoT data.

Data streaming services are just that: services. They are applications running on a server, whether virtual, physical, or container-based, that are accessed across the network. They should be secure with authentication and encryption when available. Another feature that can be useful is checkpointing, as it can aid in preventing data loss in streaming services. Checkpoint, a term used specifically in Apache Spark but a feature available in many streaming applications, stores the data in reliable storage so that you can "go back in time" to recover the data should something fail in the stream.

Given that applications must be given some level of access to the databases or streaming services or possibly directly to the IoT Data protocols, security is important here as well. Applications are accessed by users and this access is often based on usernames and passwords. Therefore, having a good password policy and enforcing it is essential. Extreme measures can be taken for critical systems, such as power grid, health and safety, etc., requiring access only from specific locations and using authentication to gain location access as well as system access. In most such cases, multi-factor authentication would be used as well.

Finally, all monitoring and maintenance systems should be secured. Particularly those that can perform automated maintenance. The last thing you want is for an attacker to downgrade all your devices through the maintenance system to a level including a vulnerability that allows them to attack the devices... well, you get the picture. Security of the maintenance system is important because it, at the very least, will provide documented information to an attacker about the systems in use. The same is true of the monitoring system While it might seem trivial at first, remember that one of the first steps in an attack is discovery. I need not scan your networks if I can simply access your monitoring dashboard and see all the details about the systems, services, and devices running on your network. Therefore, authentication to the monitoring and maintenance systems is essential

Most of the advice in this section, hopefully, is well-known and understood by anyone working in the IT space at all. However, if you want to go deeper into application level security, I highly recommend *Web Application Security: Exploitation and Countermeasures for Modern Web Applications* by Andrew Hoffman, *Security Strategies in Windows Platforms and Applications, 3rd Edition* by Michael Solomon, and *Security Strategies in Linux Platforms and Applications, Second Edition* by Michael Jang and Ric Messier.

For database security, excellent resources include *MongoDB Topology Design: Scalability, Security, and Compliance on a Global Scale* by Nicholas Cottrell, *Securing SQL Server: DBAs Defending the Database, Second Edition* by Peter A. Carter, and *Architecting Modern Data Platforms: A Guide to Enterprise Hadoop at Scale* by Kunigk, Buss, Wilkinson, and George.

7.6: Producing Design Documentation

We have now considered the many parts and pieces that must be designed or planned to have an effective wireless IoT design. The final topic of this chapter is one that should have been started long ago: design documentation. As you are designing or planning each of the components, for example the four groups in our framework, you can document the resulting decisions. If you do this throughout the process, your design documentation will be nearly complete at the moment you complete the design process.

Why is documentation important? Unless you are going to deploy every component yourself, you will need documentation to ensure that every technician deploys their part of the solution so that it works for your entire IoT solution. Even in a small IoT deployment, you are likely to have the following components:

- Wireless end devices
- Wireless gateways
- Wired network connections
- IoT Data protocol services
- Databases services
- Applications

If you must deploy each component yourself and each component averages two days to complete, it will take you between 2 and 2.5 weeks to complete the deployment. Now, imagine you have 2000 end devices, 20 gateways, 3 IoT Data protocols services, 4 database servers, and 5 applications. This would be a medium-sized IoT project, and it will likely take a single individual between 3 and 6 months to deploy it. However, with multiple individuals and teams of individuals working together, it could be deployed in 2-4 weeks (assuming no custom software development is required). The latter option is only possible if the design has moved from your head and into the documentation.

The following minimum components should be included in the design documentation:

- **Bill of Materials (BoM):** The listing of hardware and software licenses required for the deployment. It will include end devices, gateways, routers, cabling, new wired switches or routers, servers, cloud subscriptions, etc.

- **Wireless Design Report:** The collection of documents used to understand and implement the design.
    - Heat maps
    - Device placement maps
    - Cabling runs
    - Configuration parameters (including channels, transmit power, and other features)
- **Phsyical Installation Guide:** The procedures used to install the hardware components. May include wireless devices and wired devices and servers.
- **Required Supporting Services:** DHCP, DNS, NTP, Cloud, Fog, Edge, SNMP, etc.
- **IoT Data Protocol Services:** MQTT, DDS, AMQP, HTTPS, etc.
- **Data Storage Plan:** Database systems required and configuration parameters.
- **Application Plan:** Applications planned and configuration parameters.

For the wireless design report, depending on the design software you choose, you may be able to generate the entire report from the software automatically. It can then be assembled with the other document for a complete design document set. Some components are best documented in diagrams and others are best documented in lists, step-by-step procedures, and other configuration elements. If using containers for part of the solution and the containers are available online, the specific containers and versions to use should be documented and a JSON file used by an orchestration solution may be created that can simply be implemented.

7.7: Concurrent vs. Sequential Design

As a final brief topic before we finish the design chapters, consider the options of concurrent and sequential design as represented in Figure 7.21. With concurrent design, one individual or team is designing a part of the solution, while other individuals or teams are designing other parts. For example, one may design the network, while another designs the data storage and streaming, and another designs the applications. Of course, the teams must collaborate well, but concurrent design can result in a faster design process.

In sequential design, typically used because of limited technical resources, one part is designed first and then another and so on. To total design time is much longer, but one could argue that the resulting design may be more cohesive. However, with effective collaboration, the concurrent model can result in a quality, cohesive design as well.

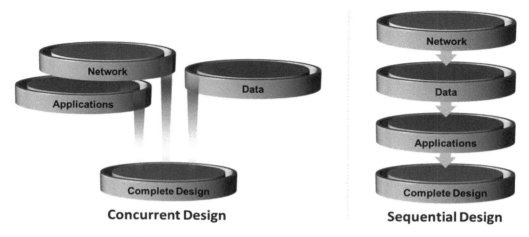

Figure 7.21: Concurrent vs. Sequential Design

7.7: Chapter Summary

In this chapter, you learned about the planning of IoT Data protocols, data storage and streaming, applications, and monitoring and maintenance. At this point, the concepts that should be addressed in design have been explored and documentation of the design should be complete. You are ready to deploy and validate the solution in Chapter 8.

7.8: Review Questions

1. What IoT data protocol uses topics as part of its defined operations?

    a. MQTT

    b. HTTP

    c. HTTPS

    d. 6LoWPAN

2. What IoT protocol uses a global data space (GDS)?

    a. AMQP

    b. MQTT-SN

    c. DDS

    d. None of these

3. True or False: MQTT should not be used on large networks because it has no QoS support.

    a. True

    b. False

4. What kind of database uses documents as the primary storage mechanism?

    a. SQL Server

    b. MySQL

    c. MongoDB

    d. Oracle

5. True or False: NoSQL databases are best for all IoT solutions.

    a. True

    b. False

6. What is not used with traditional containers?

    a. Orchestration

    b. Hypervisor

    c. Dependencies

    d. Storage

7. What is an example of a container orchestration solution?

    a. Podman

    b. LXC

    c. Windows Containers

    d. Amazon ECS

8. What kind of maintenance may be triggered by Condition-Based Preventive Maintenance?

    a. Predictive

    b. Adaptive

    c. Corrective

    d. Deferred

9. True or False: Node-RED can be used to build an IoT monitoring solution.

    a. True

    b. False

10. True or False: Python can be used to build an IoT monitoring solution.

    a. True

    b. False

7.9: Review Answers

1. **The correct answer is A.** MQTT uses topics, the other listed protocols do not.

2. **The correct answer is C.** DDS uses a GDS.

3. **The correct answer is B.** The statement is false. MQTT does support QoS.

4. **The correct answer is C.** MongoDB stores JSON documents.

5. **The correct answer is B.** The statement is false. NoSQL databases are not best for all IoT solution, though they are good for many of them.

6. **The correct answer is B.** Traditional container services do not use a hypervisor.

7. **The correct answer is D.** The Elastic Container Service (ECS) is a container orchestration service within Amazon AWS.

8. **The correct answer is A.** Predictive maintenance may be triggered by Condition-Based Preventive Maintenance because the changing condition informs of a future state.

9. **The correct answer is A.** The statement is true. Node-RED can interact with many protocols and databases and has dashboard components as well.

10. **The correct answer is A.** The statement is true. Many Python modules are available that can assist in the creation of a monitoring solution.

Chapter 8: Deploying and Validating the IoT Solution

Objectives Covered:

4.1 Validate that the RF requirements are met by the solution

4.2 Validate that the IoT solution is functioning as defined in the solution requirements

4.3 Recommend and/or perform appropriate corrective actions as needed based on validation results for RF requirements and IoT solution functionality requirements

4.4 Create a validation and test report including solution documentation and asset inventory/asset documentation

4.5 Final meeting (Q&A and hand-off)

After the IoT solution is designed and documented, deployment can begin. The CWIDP objectives do not test your knowledge of the deployment process, which is tested in CWISA, CWIIP and somewhat in CWICP. However, we will address it here before moving onto the final knowledge domain of Validate and Optimize the Wireless IoT Solution.

Validation of the solution after implementation is important. You do not want the only real validation that occurs to be your users attempting to perform actions within the system that are not functioning as required. It is far better to evaluate the system first to ensure, as much as is possible, that the system indeed meets the requirements.

Let's begin with a brief discussion of the deployment process.

8.1: Deploying the Solution

Before you can deploy the designed solution, you must achieve approval for the design. A meeting with the acquirer, key stakeholders, technical professionals within the organization, and implementors who will be assisting with the deployment should be involved. Once you have approval, or make design changes[221] and then achieve approval, you can meet with the deployment team to develop a plan for deployment. In most cases, such a plan will take a few hours and not days because time has passed since the approvals meeting and each team or individual should come to the deployment planning meeting with a plan for their components.

Sequential Deployment Plans

In a sequential deployment plan, components of the design are implemented in the most logical sequence for the design. A fully sequential plan is most often used when small teams or even an individual is responsible for performing the deployment tasks.

---

[221] It's difficult to use those four words and walk away, "or make design changes." Those four words could mean quickly adjusting a few parameters in a matter of minutes or hours or they could mean going back to the drawing board and practically starting over from scratch. The good news is that if you use the processes outlined in chapters 3-4, you are far less likely to require a complete redesign to gain approval because you will have a strong set of requirements going into the design process. If, however, you do not perform detailed requirements engineering, those four words can make for a very bad day.

However, it may be used with a large-scale design and a large deployment team if it makes the most sense for the deployment schedule.

Sometimes, normal operation schedules in an organization prohibit doing particular work at particular times. For example, the organization may only want you deploying wireless gear in the manufacturing plant on weekends. This does not mean that no work can be done. Servers can be prepared, gateways can be preconfigured, end devices can be flashed with firmware and configured, and so on. Many takes can be performed in a staging environment and then the devices can be moved to the implementation area at the right time.

Therefore, sequential deployment plans do not have to result in no work for some people and lots of work for others. The teams can perform prework that will make the actual physical deployment much more seamless.

Concurrent Deployment Plans

Concurrent deployment plans allow for different teams to work on different parts of the system concurrently. In complex deployments for large-scale solutions, this is a more common method than a purely sequential one. Like concurrent design, concurrent deployment allows for multiple deployment tasks to be performed in parallel.

While one team is working on the wireless network deployment, another team can be working on the database system deployment or preparation, and still another team can be working on application deployment or preparation. Given that the puzzle is not complete until all the puzzle pieces are present and full communications and functionality cannot be tested until they are all there, concurrent deployment general works best in these complex projects.

Solving Deployment Problems and Updating Documentation

During deployment, it is likely that some problems may be encountered. For example, a server that so and so was certain would support running the mongoDB database turns out to be overutilized already, or it's the wrong OS version, or it doesn't have a particular dependency and the dependency cannot be loaded because... you get the picture. The problems happen. All you can do is find a solution and update the documentation to reflect the new plan.

After the deployment plan is completed, or sufficiently completed to begin validating some components, we can move on to validation.

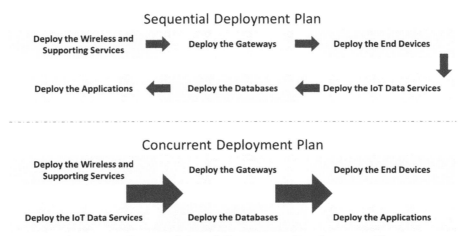

Figure 8.1: Sequential and Concurrent Deployment Plans

The SEBoK and System Realization

According to the Systems Engineering Body of Knowledge (SEBoK)[222], deployment and validation is known as *System Realization*. They divide the process of System Realization into Implementation, Integration, Verification, and Validation. Like the IEEE standards, they differentiate between verification and validation. IEEE 29148-2018 defines *verification* as confirmation, through the provision of objective evidence, that specified requirements have been fulfilled[223]. The same standard defines *validation* as confirmation, through the provision of objective evidence, that the requirements for a specific intended

---

[222] SEBoK Editorial Board. 2021. The Guide to the Systems Engineering Body of Knowledge (SEBoK), v. 2.4, R.J. Cloutier (Editor in Chief). Hoboken, NJ: The Trustees of the Stevens Institute of Technology. Accessed [DATE]. www.sebokwiki.org. BKCASE is managed and maintained by the Stevens Institute of Technology Systems Engineering Research Center, the International Council on Systems Engineering, and the Institute of Electrical and Electronics Engineers Systems Council.
[223] IEEE 29148-2018 Clause 3.1.37

use or application have been fulfilled[224]. The first determines that the system has been built right and the second determines that the right system has been built.

The SEBoK defines verification as a set of actions used to check the correctness of any element and it defines validation as a set of actions used to check the compliance of any element with its purpose and functions. So, it is in complete agreement with the IEEE standard definitions. In both cases, verification proves that the system meets the documented system or component requirements, and validation proves that the resulting system does what is needed. Stated again, verification determines that the system has been built right and validation determines that the right system has been built.

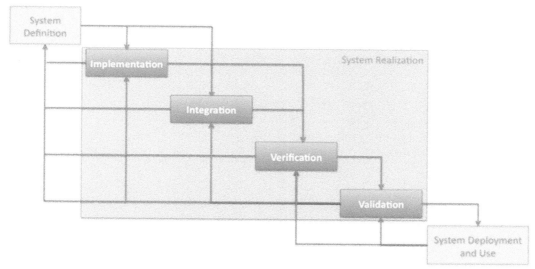

Figure 8.2: SEBoK System Realization Process[225]

The SEBoK defines implementation and integration separately though the processes may overlap. Implementation is the process of constructing the system elements that meet the requirements, while integration is taking delivery of the implemented system elements which together compose the system and aggregating them into an interconnected

---

[224] Ibid. Clause 3.1.36
[225] Ibid. SEBoK Editorial Board.

functioning whole. These four processes, implementation, integration, verification, and validation take you from the System Definition (the design based on the requirements) to a System in Deployment and Use.

Offering more options than the simple sequential and concurrent methods of deployment, the SEBoK assumes elements will be prepared individually and then integration will occur. They offer seven different models for integration as shown in Table 8.1.

Table 8.1: SEBoK Integration Techniques

Integration Technique	Description
Global Integration	Also known as *big-bang integration*; all the delivered implemented elements are assembled in only one step. - This technique is simple and does not require simulating the implemented elements not being available at that time. - Difficult to detect and localize faults; interface faults are detected late. - Should be reserved for simple systems, with few interactions and few implemented elements without technological risks.
Integration "with the Stream"	The delivered implemented elements are assembled as they become available. - Allows starting the integration quickly. - Complex to implement because of the necessity to simulate the implemented elements not yet available. Impossible to control the end-to-end "functional chains"; consequently, global tests are postponed very late in the schedule. - Should be reserved for well-known and controlled systems without technological risks.
Incremental Integration	In a predefined order, either one or a very few implemented elements are added to an already integrated increment of implemented elements. - Fast localization of faults: a new fault is usually localized in lately integrated implemented elements or dependent of a faulty interface. - Require simulators for absent implemented elements. Require many test cases, as each implemented element addition requires the verification of the new configuration and regression testing. - Applicable to any type of architecture. (Continued)

Subsets Integration	Implemented elements are assembled by subsets, and then subsets are assembled together (a subset is an aggregate); could also be called "functional chains integration".<ul><li>Time saving due to parallel integration of subsets; delivery of partial products is possible. Requires less means and fewer test cases than integration by increments.</li><li>Subsets shall be defined during the design.</li><li>Applicable to architectures composed of sub-systems.</li></ul>
Top-Down Integration	Implemented elements or aggregates are integrated in their activation or utilization order.<ul><li>Availability of a skeleton and early detection of architectural faults, definition of test cases close to reality, and the re-use of test data sets possible.</li><li>Many stubs/caps need to be created; difficult to define test cases of the leaf-implemented elements (lowest level).</li><li>Mainly used in software domain. Start from the implemented element of higher level; implemented elements of lower level are added until leaf-implemented elements.</li></ul>
Bottom-Up Integration	Implemented elements or aggregates are integrated in the opposite order of their activation or utilization.<ul><li>Easy definition of test cases; early detection of faults (usually localized in the leaf-implemented elements); reduce the number of simulators to be used. An aggregate can be a sub-system.</li><li>Test cases shall be redefined for each step, drivers are difficult to define and realize, implemented elements of lower levels are "over-tested", and does not allow architectural faults to be quickly detected.</li><li>Mainly used in software domain, but can be used in any kind of system.</li></ul>
Criterion Driven Integration	The most critical implemented elements compared to the selected criterion are first integrated (dependability, complexity, technological innovation, etc.). Criteria are generally related to risks.<ul><li>Allows early and intensive testing of critical implemented elements; early verification of design choices.</li><li>Test cases and test data sets are difficult to define.</li></ul>

Table 8.1: SEBoK Integration Techniques[226]

---

[226] Ibid. SEBoK Editorial Board.

Figure 8.3[227] shows the processes as defined by the Defense Acquisition University (DAU) and shared in the SEBoK for the planning, design, and deployment of a system within the systems engineering process mapped to the CWNP wireless IoT solution design phases. You will notice the overlaps that exist. You can, for example, begin implementation and integration at some levels even while requirements engineering, and architecture design are still in process. In fact, verification can also begin before full implementation is complete, and transition can phase in as well.

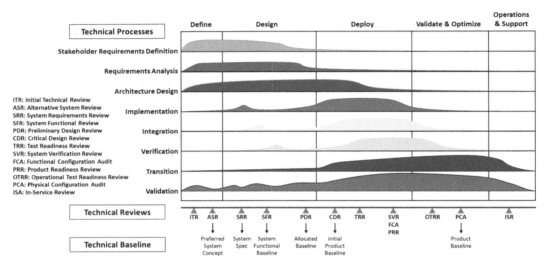

Figure 8.3: Systems Engineering Technical Processes Mapped to the CWNP Phases

Given that we are focused on deployment (implementation and integration), note the overlap in these areas in Figure 8.3. Indeed, when you start implementing the various components within the system, there is a point where you can begin integrating them as well. When an entire component or element is implemented, you can also begin verification. Finally, notice that validation is performed all throughout the processes, though it becomes much more intense towards the end. Certainly, the validation

---

[227] The Defense Acquisition University (DAU) released the process diagram recreated in Figure 8.3 into the public domain in 2010 and I'm glad they did. It shows the same basic concept we discussed earlier in this book that I have been explaining for 20 years. Interestingly, and to be completely open, I came across it quite by accident while reviewing material from the SEBoK during the writing of this chapter.

processes we're about to discuss, need not wait until deployment is complete. If others are deploying the designed solution and you will perform the validation, you can begin the processes as soon as functional components requiring validation are available.

8.2: Validate RF Requirements

Though you may have toured the facilities before the design of the network, it is always amazing how some installations can turn out to be bad. Whether it is a pipe for sprinklers which was not on the plan or a gateway installer which felt creative in the way he was mounting the gateway or orienting the antennas, a simple walkthrough of the site to check that the installation was performed to your standards will save you valuable time when you survey. Stated differently, break out your eyes before you break out your tools. Visually inspect the deployment and repair or resolve any installation location problems related to the RF equipment.

As an example, an expert walking through an area and noticing an end device nestled between three large pieces of metal enclosed machinery in a space less than one square meter, though it need not be there, knows that the device is likely to experience problems in communication. Such an issue can be quickly resolved through visual inspection.

Simple observation can sometimes give you a good indication of the checks you may need to do once this walk-through is done and many issues can be quickly repaired on-the-fly. If the LEDs on the gateways were left visible, you could refer to your vendor's documentation to see if they are properly powered, have connected to the Ethernet or other backhaul network, and have active radios. Most vendors can even indicate through the led patterns whether end devices are connected or if there are irregularities with the gateway, such as power cycling, flash corruption and the like. Refer to your vendor's documentation for more information on device-specific patterns.

Next you can break out the RF tools and validate coverage requirements in all areas where end devices must function. However, this is a point of interest in IoT networks. If the end devices are mostly stationary, simply validating that they are transmitting data through the gateway to the network tells you that they have coverage. Now, if you are only seeing fifty percent of the traffic that you should be seeing from an end device, it tells you there may be either a coverage or interference problem where it is located, and you can resolve it. The point is that, given the nature of IoT devices and their common communication patterns, you can often validate coverage and even capacity by simply evaluating the communications coming into the network.

Like coverage, the simplest way to verify that capacity requirements have been met is to determine if the devices are achieving an acceptable rate of transmission per the requirements. For example, if the requirements state that 70 percent of end device transmissions must be successfully delivered, which is actually not a low amount for some IoT solutions (some tolerate up to 50 percent failed transmissions), and you know that end devices should be transmitting 100 messages per hour if one hundred percent are delivered, you can simply evaluate whether the average messages from end devices is above 70 per hour. If so, in general, capacity requirements have been met. However, the closer the average is to 70, the more important it will be to inspect the lowest performing end devices to bring them up to capacity requirements.

Neither of these recommendations is intended to diminish the use of RF analysis tools to verify coverage and capacity; however, the analysis of capacity is quite challenging in IoT solutions, unless you've implemented a monitoring solution that can alert you to poorly performing end devices and inform you of network performance statistics. Few "walkabout" protocol analyzers exist that provide you with capacity analysis (in a simple view) for most wireless IoT protocols.

You can also use the solution to validate the solution in reverse for actuators. Send commands to the actuators, are they working? This method is bi-directional. The point is to take advantage of the whole purpose of the system to test the system. The purpose of the system is for things to communicate at some level above the wireless network (in most cases). Are the communications reaching that level and are communications from that level reaching the things.

In addition to the validation of coverage and capacity, you should look for sources of interference and resolve them. It may have been several months since the site visit that started this process and changes could have occurred in the environment. A spectrum scan of the space, while walking through it, can be used to identify and then locate any sources of interference. The basic process is as follows:

1. Walk through the environment slowly while monitoring the spectrum analyzer views.

2. If you notice a signal out-of-character with the IoT protocols implemented and in the same frequency ranges in use and of sufficient duty cycle to cause problems, document that interferer and then seek it out.
3. Move about until the signal is the strongest and then visually inspect the area to locate and identify the source.
4. If it is a new interferer and not one for which the design has already accommodated, document the source and a recommendation for resolution.
5. Repeat 1-4 until the entire facility has been explored.

In large environments, this can take several hours; however, it is unlikely to take days because you are not likely to find a large number of new interferers that were not there in the beginning of the project. Notice that we are not redocumenting known and existing interferers. They should have been accounted for already in the design.

8.3: Validate IoT Solution Requirements

The next step is to validate that the IoT solution is functioning as defined in the solution requirements. In the preceding step, you validated the RF requirements were met and so the wireless network should be able to function as intended. However, this does not mean that it is functioning as intended. Gateways could be misconfigured. The wrong antennas could have been used or any number of other problems, including higher layer problems, could be preventing the solution from function. We need to validate that end-to-end functionality is operating as required. We will work our way through these validations starting from visual inspection and ending in active communication tests.

Verify Aesthetics Requirements

If aesthetics requirements were specified in the requirements engineering process, we must ensure that they have been met. This task is achieved by visually inspecting the installation of end devices, gateways, and any other components that may be visible within the operational sphere of the organization's workspaces. If you encounter an aesthetics violation, document it, and form a recommendation for resolution. Resolution options may include:

- Reskinning the device so that it blends with the environment.
- Mounting devices out of view as long as it does not hinder communications.

- Using special deployment constructs that are designed to allow for RF propagation with very little attenuation or reflection but are aesthetically pleasing in the environment.

Verify Power and Grounding Requirements

Another validation task is the power and grounding requirements. Power validation can often be validated by inspecting LEDs on devices and ensuring that they report proper power. If devices are powered through PoE, such as gateways, the PoE used by the switch can be inspected to verify proper power is being delivered.

Grounding requirements can often be validated through visual inspection. Additionally, you may use a ground resistance tester. Grounding helps to protect against lighting strike incidental energy, but for outdoor mounted antennas, lighting arrestors should have been deployed as well and this can be verified visually.

To perform advanced grounding tests, you can also perform a Wenner method test. This test is performed by inserting four evenly spaced electrodes in the ground. The two outer electrodes inject current into the soil. The two inner electrodes measure voltage. With this configuration and a ground testing tool, you can determine the resistivity of the soil. The soil resistivity determines the effectiveness of a grounding system. Doubling the distance the ground rod is driven into the ground reduces resistance by 40% up to certain limits.

Verify Channel Selections and Transmit Power

The next step is to validate that the wireless network is properly configured per the design specification based on channels selected and transmit power. Channels in use can be identified with a spectrum analyzer if the resolution is high enough and transmissions are active. If not, manually viewing the configuration of the gateways can achieve this result.

Transmit power validation is nearly impossible with portable spectrum analyzers because the ability to identify the originating transmit power is too inaccurate. You can certainly verify that it not below a particular level but determining exactly how much transmit power is used will be a challenge with such tools. Instead, like the channels, verify the configuration in the gateways.

This task can be very time consuming if you have dozens of gateways and no centralized configuration solution, which is not uncommon with many IoT solutions. You may choose to provide step-by-step instructions and recruit a team to help you with the process.

Verify Proper Security

Validating security configuration is an end-to-end process. Depending on the IoT solution and your role, you may need to validate the following:

- **Encryption:** wireless link encryption, wired network encryption, data storage encryption, encrypted Internet tunnels, and encryption in cloud solutions.
- **Authentication:** authentication to the wireless network, authentication between end devices and applications, authentication between gateways and applications, authentication between end devices and IoT Data protocol services, authentication between applications and IoT Data protocol services, authentication between applications and databases, authentication to the cloud, authentication to applications and databases in the cloud.
- **Accounting/Logging:** verify that logging is implement in every area of the solution where it is required.
- **Physical Security:** verify that administration interfaces on devices are disabled, removed, or blocked; verify that devices are securely mounted; verify that devices have proper enclosures for the environment.

Conduct Device Testing

Only after performing the preceding validation tests should you begin device testing. There is not reason to test an end device and its operations until you've verified that everything is properly in place that should be in place for functionality.

Device testing may not be performed with every end device, but it should be performed with every class of device. For example, you may have deployed 30 of one end device, 45 of another, 123 of another, and 80 of another. Testing one or two samples from each class should be sufficient to ensure the following:

- When the device is power cycled it returns to a proper operational state.
- The device can establish and maintain a connection to the network.

- The device is operating at an acceptable temperature in the environment.
- The device data is reaching the IoT Data protocol services, applications, cloud, and/or databases.

Some level of trust must be placed in the deployment team, particularly if thousands of devices have been installed. You certainly cannot test every one of the devices. Instead, test some sample devices and then gather metrics of reported data on the network, ensure that all devices are reporting data, and any devices that have not reported data should be further evaluated to determine why.

One final note on device testing. It might be tempting to thinks that power consumption testing for battery powered devices should be performed at this stage. However, that is not the case. This type of testing should have been performed during the design phase when selecting the end devices. If you have already purchased hundreds or thousands of devices and they are overconsuming battery power to the point where battery life will be significantly less than desired, you are dealing with a problem that should never be dealt with at this stage. Always perform power consumption testing on battery powered end devices before purchasing in volume[228].

Conduct Mobility Testing

Not all devices require mobility testing. This should be obvious. Only those that are mobile require it. Given that you've already validated coverage during the RF tests, it is likely that most mobile devices will work fine. However, they should be tested with actual mobility to ensure that the acceptable percentage of messages are getting through to the network while mobile and that connectivity is always available where it is required.

Again, due to the nature of IoT, the simplest way to test mobility is to work with a partner. You or the partner can monitor the incoming messages from the IoT device and

---

[228] This is an important area of difference in wireless IoT design and deployment when compared to Wi-Fi deployments, at least, common Wi-Fi deployments. Most Wi-Fi deployments are about user-based end devices like laptops, tablets, and mobile phones. In such deployments, these end devices can be recharged easily enough and, while it may be frustrating if the battery power doesn't last as long as the user would like, it doesn't necessarily break the network. With IoT end devices, having insufficient battery power will indeed break the network. Therefore, testing a key element such as this should happen early in the system life cycle.

the other can move the device at the normal rate of movement during normal operations and throughout all the areas where it must function. If the device is transmitting the expected messages throughout this time and during this activity and they are arriving at the application level, the solution is working.

Some devices transmit intermittently making this more challenging to perform. If the device can be reconfigured to communicate at an interval or has a link testing facility, you may choose to use that to simplify the process[229].

After these tests are complete, you can move onto the recommendation phase where you will either recommend and perform problem resolutions or you will recommend and document them for others to perform.

8.4: Recommend or Perform Corrective Actions in a Validation and Test Report

The next step, after validation of the RF and the IoT solution requirements, is to recommend or perform appropriate corrective actions as needed based on the validation results for RF and the IoT solution functionality. This can be done informally or in a documented report. If you are doing the corrective work, it may be an informal conversation with light documentation of the actions you will take. If you are simply providing guidance on the tasks that should be performed, this guidance can be documented in the validation and test report. We have combined these two exam objectives together but know that they are sometimes separate processes.

The documentation for corrective actions will list the problems discovered and the recommended solutions. It is usually best to categorize them into groups for readability and provide the optimal sequence for resolution. For example, resolving problems in the higher layers should usually be performed before resolving problems in the lower layers.

---

[229] If you have chosen to use cellular IoT devices for your mobility needs, you will be able to perform similar tests. The devices will still communicate as they do, they are just using the cellular network for the communication. You can monitor the cloud side and verify that the device is consistently transmitting as expected. This cloud-side or server-side or IoT Data protocol-side testing is the simplest and often most effect method for testing both mobility and general device functional operations.

Repair the issues from the top down. The reason for this is simple: if the top is not working, you cannot validate that the bottom is working properly when those issues are resolved.

The validation and test report may include the recommended corrective actions as well as the final report. This test report will typically include the original requirements with a statement of validation that they have been achieved. It will also convert the BoM into an asset inventory that will include asset documentation, such as general or specific device locations, device IDs may be included, or they may be in a separate document.

Given that organizations struggle to maintain documentation. An excellent "value add" that can be provided at the end of the project is a complete system documentation set that provides architecture diagrams, network diagrams, devices and configurations, and application descriptions and dependencies. Much of this information will already be in the design documents and it will simply involve adjusting it based on the actual implementation and any required changed discovered and made during validation.

8.5: Final Meeting (Q&A and Hand-Off)

The final meeting is the closure point of the project and when the system officially completes the transition from development (define, design, deploy, and optimize & validate) to operations.

Provide an opportunity for the acquirer and stakeholders to ask questions as well as the operations team that will manage the system. Ensure that all questions are answered, and the answers are understood. Verify that everyone feels confident that they have what they need to succeed with the new system and, if so, with final approval of the project, move on to the next one.

8.6: Chapter Summary

In this final chapter, you explored the deployment plans common for deployment of a system design. Then you learned about the validation and optimization tests and procedures required to finalize the deployment.

Congratulations! You've built a system from nothing to the Internet of Everything.

8.7: Review Questions

1) What kind of deployment plan involves completing each task before moving on to the next one?
   a) Sequential
   b) Concurrent
   c) Big Bang
   d) None of these

2) In the SEBoK System Realization process, what two components involve the work tasks of deploying a solution?
   a) Implementation and Irrigation
   b) Verification and Planning
   c) Validation and Planning
   d) Implementation and Integration

3) True or False: According to the DAU model for technical processes, validation is often performed throughout the entire process.
   a) True
   b) False

4) What can be used to verify that radios are operational in a gateway?
   a) Antenna orientation
   b) LEDs
   c) Mounting location
   d) None of These

5) True or False: Proper IoT device operation can only be validated with a protocol analyzer that supports the protocol in use.
   a) True
   b) False

6) When interferers are detected during validation, which interferes should be a concern?
   a) All interferers
   b) Only interferers in 2.4 GHz
   c) New interferers not seen during the initial site visit
   d) None of these

7) What is a practical solution to aesthetics problems related to a gateway?
   a) Place a plant over the gateway
   b) Mount the gateway out of view
   c) Remove the gateway
   d) Reconfigure the gateway for an invisible channel

8) True or False: The Wenner method test is used to verify coverage in wireless networks.
   a) True
   b) False

9) Which industry vertical has the most aesthetic concerns?
   a) Retail
   b) Hospitality
   c) Healthcare
   d) Educational

10) Which requirements should be reported as validated in the validation and test report?
    a) All of them
    b) Only stakeholder requirements
    c) Only technical requirements
    d) None of them

8.8: Review Answers

1) The correct answer is A. Sequential plans work by completing one component before moving onto the next.
2) The correct answer is D. Implementation and integration.
3) The correct answer is A. The statement is true. Validation can be performed throughout the entire process.
4) The correct answer is B. The LEDs may indicate radio status.
5) The correct answer is B. The statement is false because proper IoT device operation can be validated without a protocol analyzer.
6) The correct answer is C. New interferers are a concern. Those identified in the initial site visit should have been addressed in the design.
7) The correct answer is B. While not always the best solution, mounting the gateway out of view is a practical solution.
8) The correct answer is B. The statement is false. The Wenner method is used to determining the soil resistivity for grounding purposes.
9) The correct answer is B. The hospitality industry is the vertical best known for such concerns.
10) The correct answer is A. All requirements should be reported as validated because they should have been validated.

NOTES

NOTES

NOTES

NOTES

NOTES

NOTES

NOTES

NOTES

NOTES

NOTES

CPSIA information can be obtained
at www.ICGtesting.com
Printed in the USA
LVHW021111291021
701899LV00002B/3